Vivek Bannore

Iterative-Interpolation Super-Resolution Image Reconstruction

Studies in Computational Intelligence, Volume 195

Editor-in-Chief
Prof. Janusz Kacprzyk
Systems Research Institute
Polish Academy of Sciences
ul. Newelska 6
01-447 Warsaw
Poland
E-mail: kacprzyk@ibspan.waw.pl

Further volumes of this series can be found on our homepage: springer.com

Vol. 172. I-Hsien Ting and Hui-Ju Wu (Eds.)
Web Mining Applications in E-Commerce and E-Services, 2009
ISBN 978-3-540-88080-6

Vol. 173. Tobias Grosche
Computational Intelligence in Integrated Airline Scheduling, 2009
ISBN 978-3-540-89886-3

Vol. 174. Ajith Abraham, Rafael Falcón and Rafael Bello (Eds.)
Rough Set Theory: A True Landmark in Data Analysis, 2009
ISBN 978-3-540-89886-3

Vol. 175. Godfrey C. Onwubolu and Donald Davendra (Eds.)
Differential Evolution: A Handbook for Global Permutation-Based Combinatorial Optimization, 2009
ISBN 978-3-540-92150-9

Vol. 176. Beniamino Murgante, Giuseppe Borruso and Alessandra Lapucci (Eds.)
Geocomputation and Urban Planning, 2009
ISBN 978-3-540-89929-7

Vol. 177. Dikai Liu, Lingfeng Wang and Kay Chen Tan (Eds.)
Design and Control of Intelligent Robotic Systems, 2009
ISBN 978-3-540-89932-7

Vol. 178. Swagatam Das, Ajith Abraham and Amit Konar
Metaheuristic Clustering, 2009
ISBN 978-3-540-92172-1

Vol. 179. Mircea Gh. Negoita and Sorin Hintea
Bio-Inspired Technologies for the Hardware of Adaptive Systems, 2009
ISBN 978-3-540-76994-1

Vol. 180. Wojciech Mitkowski and Janusz Kacprzyk (Eds.)
Modelling Dynamics in Processes and Systems, 2009
ISBN 978-3-540-92202-5

Vol. 181. Georgios Miaoulis and Dimitri Plemenos (Eds.)
Intelligent Scene Modelling Information Systems, 2009
ISBN 978-3-540-92901-7

Vol. 182. Andrzej Bargiela and Witold Pedrycz (Eds.)
Human-Centric Information Processing Through Granular Modelling, 2009
ISBN 978-3-540-92915-4

Vol. 183. Marco A.C. Pacheco and Marley M.B.R. Vellasco (Eds.)
Intelligent Systems in Oil Field Development under Uncertainty, 2009
ISBN 978-3-540-92999-4

Vol. 184. Ljupco Kocarev, Zbigniew Galias and Shiguo Lian (Eds.)
Intelligent Computing Based on Chaos, 2009
ISBN 978-3-540-95971-7

Vol. 185. Anthony Brabazon and Michael O'Neill (Eds.)
Natural Computing in Computational Finance, 2009
ISBN 978-3-540-95973-1

Vol. 186. Chi-Keong Goh and Kay Chen Tan
Evolutionary Multi-objective Optimization in Uncertain Environments, 2009
ISBN 978-3-540-95975-5

Vol. 187. Mitsuo Gen, David Green, Osamu Katai, Bob McKay, Akira Namatame, Ruhul A. Sarker and Byoung-Tak Zhang (Eds.)
Intelligent and Evolutionary Systems, 2009
ISBN 978-3-540-95977-9

Vol. 188. Agustín Gutiérrez and Santiago Marco (Eds.)
Biologically Inspired Signal Processing for Chemical Sensing, 2009
ISBN 978-3-642-00175-8

Vol. 189. Sally McClean, Peter Millard, Elia El-Darzi and Chris Nugent (Eds.)
Intelligent Patient Management, 2009
ISBN 978-3-642-00178-9

Vol. 190. K.R. Venugopal, K.G. Srinivasa and L.M. Patnaik
Soft Computing for Data Mining Applications, 2009
ISBN 978-3-642-00192-5

Vol. 191. Zong Woo Geem (Ed.)
Music-Inspired Harmony Search Algorithm, 2009
ISBN 978-3-642-00184-0

Vol. 192. Agus Budiyono, Bambang Riyanto and Endra Joelianto (Eds.)
Intelligent Unmanned Systems: Theory and Applications, 2009
ISBN 978-3-642-00263-2

Vol. 193. Raymond Chiong (Ed.)
Nature-Inspired Algorithms for Optimisation, 2009
ISBN 978-3-642-00266-3

Vol. 194. Ian Dempsey, Michael O'Neill and Anthony Brabazon (Eds.)
Foundations in Grammatical Evolution for Dynamic Environments, 2009
ISBN 978-3-642-00313-4

Vol. 195. Vivek Bannore
Iterative-Interpolation Super-Resolution Image Reconstruction:
A Computationally Efficient Technique, 2009
ISBN 978-3-642-00384-4

Vivek Bannore

Iterative-Interpolation Super-Resolution Image Reconstruction

A Computationally Efficient Technique

Dr. Vivek Bannore
SCT-Building
University of South Australia
Adelaide
Mawson Lakes Campus
South Australia, SA 5095
Australia
E-mail: Vivek.Bannore@gmail.com

ISBN 978-3-642-00384-4 e-ISBN 978-3-642-00385-1

DOI 10.1007/978-3-642-00385-1

Studies in Computational Intelligence ISSN 1860949X

Library of Congress Control Number: 2009921140

Typeset & Cover Design: Scientific Publishing Services Pvt. Ltd., Chennai, India.

Printed in acid-free paper

9 8 7 6 5 4 3 2 1

springer.com

To my wife, Mitu

- Vivek Bannore

Preface

In many imaging systems, under-sampling and aliasing occurs frequently leading to degradation of image quality. Due to the limited number of sensors available on the digital cameras, the quality of images captured is also limited. Factors such as optical or atmospheric blur and sensor noise can also contribute further to the degradation of image quality. Super-Resolution is an image reconstruction technique that enhances a sequence of low-resolution images or video frames by increasing the spatial resolution of the images. Each of these low-resolution images contain only incomplete scene information and are geometrically warped, aliased, and under-sampled. Super-resolution technique intelligently fuses the incomplete scene information from several consecutive low-resolution frames to reconstruct a high-resolution representation of the original scene.

In the last decade, with the advent of new technologies in both civil and military domain, more computer vision applications are being developed with a demand for high-quality high-resolution images. In fact, the demand for higher-resolution images is exponentially increasing and the camera manufacturing technology is unable to cope up due to cost efficiency and other practical reasons. Therefore, for an imaging system, super-resolution overcomes the limitation in terms of image quality by enhancing the spatial resolution to generate high-resolution images of the original scene, without the need of any hardware enhancements. This is why most of the research into image resolution enhancement (or super-resolution) has been majorly directed towards developing techniques that deliver the highest possible fidelity of the reconstruction process. The computational efficiency issues and the feasibility of developing realistic applications based on super-resolution algorithms have attracted much less attention.

In the field of image processing, it is a widely known fact that, super-resolution is an inverse problem which is also ill-conditioned, making the estimation process highly computationally expensive. Also, the number of unknown variables to be estimated is equal to the number of pixels in the reconstructed high-resolution image, which is of the order of hundreds of thousands. Clearly, the fidelity of the reconstruction has to be traded-off by performance. It is, thus, desirable to develop algorithms that maintain the proper balance between computational performance and the fidelity of the reconstruction.

In this research, a novel and innovative, hybrid reconstruction scheme has been proposed for super-resolution reconstruction which addresses the issue of maintaining an optimum balance between improving spatial resolution and keeping the computational time low. The proposed fast reconstruction scheme is referred to as ***Iterative-Interpolation Super-Resolution*** (**IISR**). It is based on interpolation and subsequent iterative improvements of the reconstructed image to generate a high-resolution image from a sequence of geometrically warped, aliased and under-sampled, low-resolution frames. A mathematical basis for the proposed scheme is also provided. The IISR technique utilizes a relatively small number of low-resolution images, as low as 10, to generate the high-resolution image and is relatively inexpensive computationally.

The common feature of inverse problems is their sensitivity to even small perturbations of the data that may introduce significant errors in the reconstruction process. Due to this many existing super-resolution techniques are based on the optimization approach where a regularization term is added so as to stabilize the inverse problem and generate a unique solution. In this research, the optimization approach to super-resolution is re-visited. The estimation problem of super-resolution is reformulated in terms of Tikhonov regularized optimization procedure and the conjugate gradient method is adopted for finding the minimum of the resulting objective function. Also, the significance and influence of the regularization term on the accuracy of super-resolution image reconstruction is reinvestigated. An added benefit of this reinvestigation helps in comparing the performance, in terms of effectiveness and accuracy, of the IISR method with the best possible regularized optimization technique.

It is also shown that by combining the IISR technique with the regularized optimization approach, the convergence rate of the reconstruction process in the optimization approach can be significantly improved. It is also validated that the proposed IISR reconstruction technique is accurate and the further improvement by the optimization procedure is relatively small, computationally expensive and therefore impractical.

Image registration forms a significant intermediate process in achieving accurate super- resolution from a sequence of aliased and under-sampled low-resolution frames. Even though image registration is not the primary subject of this research, the significance and advantages of implementing a computationally efficient and accurate hierarchical-based image registration technique is also discussed.

Two major software frameworks have been developed based on the research presented in this research. Both softwares serve the purpose of comparative experimental tools but can also be used as standalone super-resolution image reconstruction systems.

The research is accompanied by experimental simulations and numerical results generated on synthetic and real imagery.

Thanks to Dr. Noel Martin for his support during the course of research. We would like to thank University of South Australia and Australian Defence Science and Technology Organisation for funding and supporting this project.

We would also like to express our sincere gratitude to the School of Electrical & Information Engineering and the Knowledge-based Intelligent Engineering Systems for providing an excellent research environment.

Adelaide, Australia
January 2009

Vivek Bannore

Contents

Chapter 1
Introduction to Super-Resolution

1.1 Introduction

In the last decade the world has seen an immense global advancement in technology, both in hardware and software. The industries took advantage of the advanced technology to produce electronic gadgets such as computers, mobile phones, PDAs and many more at affordable prices. The camera sensor manufacturing units also advanced in their manufacturing techniques to produce good quality high-resolution (HR) digital cameras. Although, HR digital cameras are available, many computer vision applications such as satellite imaging, target detection, medical imaging and many more still had a strong requisition for higher resolution imagery which very often exceeded the capabilities of these HR digital cameras. To cope up with the strong demand for higher-resolution imagery, these applications approached image-processing techniques for a solution to generate good quality HR imagery.

Super-Resolution image reconstruction is a promising technique of digital imaging which attempts to reconstruct HR imagery by fusing the partial information contained within a number of under-sampled low-resolution (LR) images of that scene during the image reconstruction process. Super-resolution image reconstruction involves up-sampling of under-sampled images thereby filtering out distortions such as noise and blur. In comparison to various image enhancement techniques, super-resolution image reconstruction technique not only improves the quality of under-sampled, low-resolution images by increasing their spatial resolution but also attempts to filter out distortions.

1.2 What Is Image Resolution?

As per [1] optical resolution is a measure of the ability of a camera system, or a component of a camera system, to depict picture detail. On the other hand, image resolution is defined as the fineness of detail that can be clearly distinguished in an image. Both the definitions apply to digital and analogue camera systems and images. However, in this research, the term resolution will only relate to digital camera systems and digital images. There are two most common classifications of digital image resolution, namely – spatial and bit-depth.

- Spatial resolution refers to the level of detail discernable in an image.
- Bit-depth refers to the number of bits or 0's and 1's that can be used to specify the colour at each pixel of an image.

V. Bannore: Iterative-Interpolation Super-Resolution Img. Recons., SCI 195, pp. 1 – 8.
springerlink.com

Spatial resolution essentially describes the total number of pixels in an image, horizontally and vertically. For example, a digital image 300 pixel (wide) x 300 pixel (high) consists of a total of 90,000 pixels or is a 0.1 megapixel (MP) image. If this image is tripled, the dimensions will be 900 pixels (wide) x 900 pixels (high) with a total of 810,000 pixels or 0.8 MP. Clearly, the detail carrying capacity of an image is directly proportional to the number of pixels in an image. Higher the number of pixels, higher is the detail representation of the image. On the other hand, bit-depth describes the number of possible colours at each pixel. Bit-depth is also known as colour-depth. More the bits per pixel, greater are the colours at each pixel thereby increasing the colour details in the digital image. For example, a grayscale image carries 8-bits per pixel, which in turn means that a grayscale image can have 2^8 or 256 shades of gray. For colour images, if 8-bits per pixel per channel are used, then the colour digital image can have 2^{8*3} or 16,777,216 different colours. This is also known as 24-bits per pixel.

In this research, the word resolution refers to spatial resolution unless otherwise mentioned.

1.3 Image Degradation Factors

The acquired image usually represents the scene in an unsatisfactory manner. Since real imaging systems as well as imaging conditions are imperfect, an observed image represents only a degraded version of the original scene. These degradations in the images are caused due to various factors such as blur, noise and aliasing. Figure 1.1 shows an example of a corrupted image (left-panel) and the original scene (right-panel). Such distortions may get introduced into an imaging system due to the following reasons:

- Motion between the camera sensor and the scene or subject.
- Camera optics and lenses.
- Atmosphere.
- Insufficient sampling.

Blur can be introduced into the image during the imaging process by factors such as motion of the scene, wrong focus, atmospheric turbulence and optical point spread function. To remove the effect of blurring on an image is known as de-blurring which is a well known image enhancement technique. If the imaging conditions at the time of acquiring an image are known, it is much easier to de-blur the image accurately. Figure 1.2 shows an example of a blurred image (left-panel) and the original scene (right-panel).

Noise is a random background event and is certainly not a part of the ideal scene/signal and may be caused by a wide range of sources such as variations in the detector sensitivity, optical imperfections and environmental changes. Although many noise models exist in literature, we only consider the additive Gaussian white noise since it provides a good model for noise in most of the imaging systems. The noise is also assumed to be spatially uncorrelated with respect to the image, that is, there is no correlation between the image pixel values

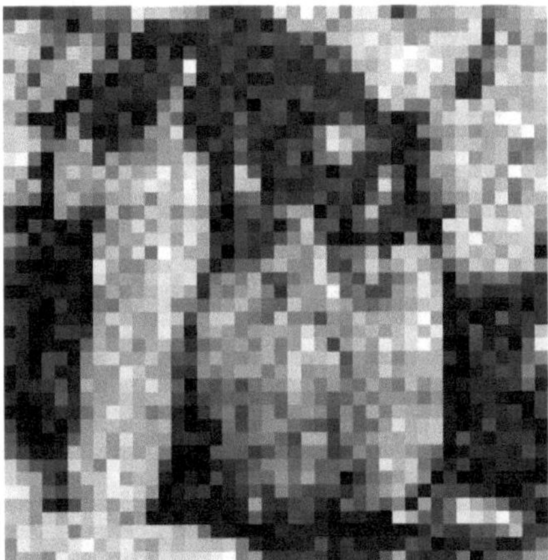

Fig. 1.1 An example of noisy, blurred and under-sampled image (left-panel) of the original scene (right-panel)

Fig. 1.2 An example of the blurred image (left-panel) of an original scene (right-panel)

and the noise components. Figure 1.3 shows an example of a noise-corrupted image (left-panel) and the original scene (right-panel).

Another factor affecting image resolution is due to the insufficient spatial sampling of the images. As per the Shannon-Nyquist Sampling theorem [2-4], the sampling frequency should be greater than twice the highest frequency of the input signal. If the sampling frequency is less than twice the highest frequency, then all frequency components higher than half the sampling frequency are reflected as lower frequencies in the reconstructed signal. This is referred to as under-sampling of images which occurs in many imaging sensors. Due to under-sampling, the high-frequency components overlap with the low-frequency components and get introduced into the reconstructed image/signal causing degradation of the image.

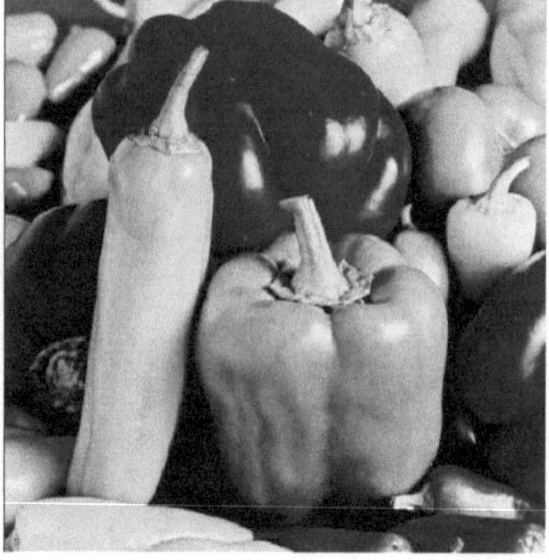

Fig. 1.3 An example of the noise-corrupted image (left-panel) of an original scene (right-panel)

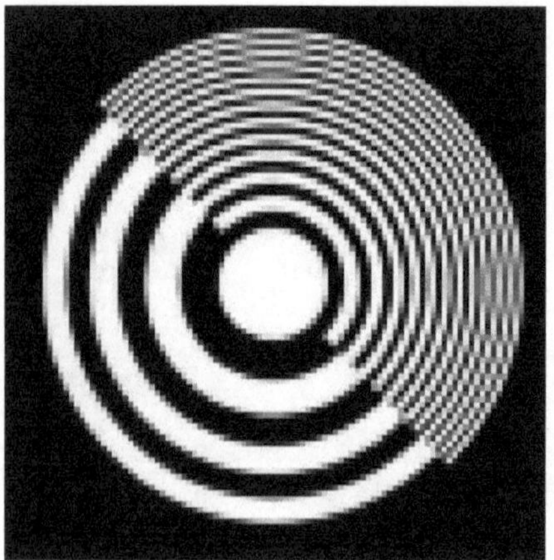
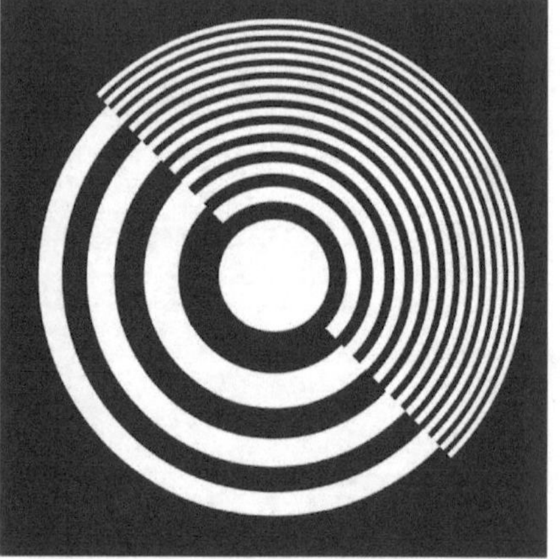

Fig. 1.4 An example of an aliased, under-sampled image (left-panel) of an original scene (right-panel)

Such degradation is known as aliasing which consequently causes partial loss of scene information. The aliasing effect may also give rise to artifacts thereby corrupting the reconstructed image. To reduce these artifacts, anti-aliasing techniques are implemented. Figure 1.4 shows an example of an aliased, under-sampled image (left-panel) and the original scene (right-panel). The moiré patterns are easily visible in fig. 1.4.

1.4 Significance of Super-Resolution

As defined earlier, spatial resolution refers to the spacing of pixels in a digital image. Therefore, more the number of pixels, more detailed is the information contained within the image. Considering this fact, with the help of advanced sensor technology, industries started manufacturing sensors with an increased number of pixels per unit area by decreasing the pixel size. But there is a limitation over the reduction of pixel size due to the shot-noise effect on the sensor. The optimum limit on the pixel size is 40 μm^2 which the current technology has already reached. In certain defence computer vision applications, unmanned aerial vehicles (UAVs) are used for acquiring images. All UAVs have a payload carrying capacity and is usually half the UAVs take-off or launch weight. It is therefore not feasible to mount heavy HR cameras with image stabilization equipment to counteract the vibrations of UAVs. The ever demanding need for high-resolution imagery stimulated research and development of super-resolution techniques.

The process of taking a sequence of under-sampled LR images of a particular scene and generating its high quality HR image is known as *Super-Resolution* image reconstruction. Due to the relative motion between the camera sensor and scene, each LR frame acquired, contains slightly different information about the scene. Super-resolution takes advantage of this distinct information contained in each under-sampled LR image and fuses the information from all the LR frames during the reconstruction process to generate a high quality HR image of the true scene.

Figure 1.5 shows a simple illustration to explain the concept of super-resolution. As per the Nyquist theorem, a signal can be perfectly reconstructed if it is sampled at a frequency twice the highest frequency of the input signal. Figure 1.5(a) represents a scene captured by a camera with the coloured dots representing the sampled points. By shifting the camera, we acquire three more frames shown by fig. 1.5(b)-to-(d). Each of the four frames contains slightly different information. These frames when fused together generate an image with higher-resolution containing information from all the four frames. This is super-resolution in its simplest form. The images are assumed to be band-limited. Super-resolution is a well-known inverse problem, the solution of which is highly unstable; the process is ill-conditioned and highly computationally expensive.

To extract complete information from multiple LR frames, it is very crucial to detect the shifts (in a global sense) between the LR frames to align them appropriately and accurately. This pre-processing step is known as *Image Registration*. It is the process of matching two or more images taken at different times, from different viewpoints and/or by different sensors. A mis-registration of frames can cause loss of important information which is why registration is very crucial for generating SR images. Registration is a huge research domain and has its use in many applications such as medical imaging, remote sensing, automatic target detection and many more.

1.5 Applications of Super-Resolution

The field of super-resolution has a vast area of application. Although the concept of super-resolution remains the same, the techniques of achieving HR imagery may or may not be the same for each and every application. This is because in certain applications such as real time video surveillance or target detection,

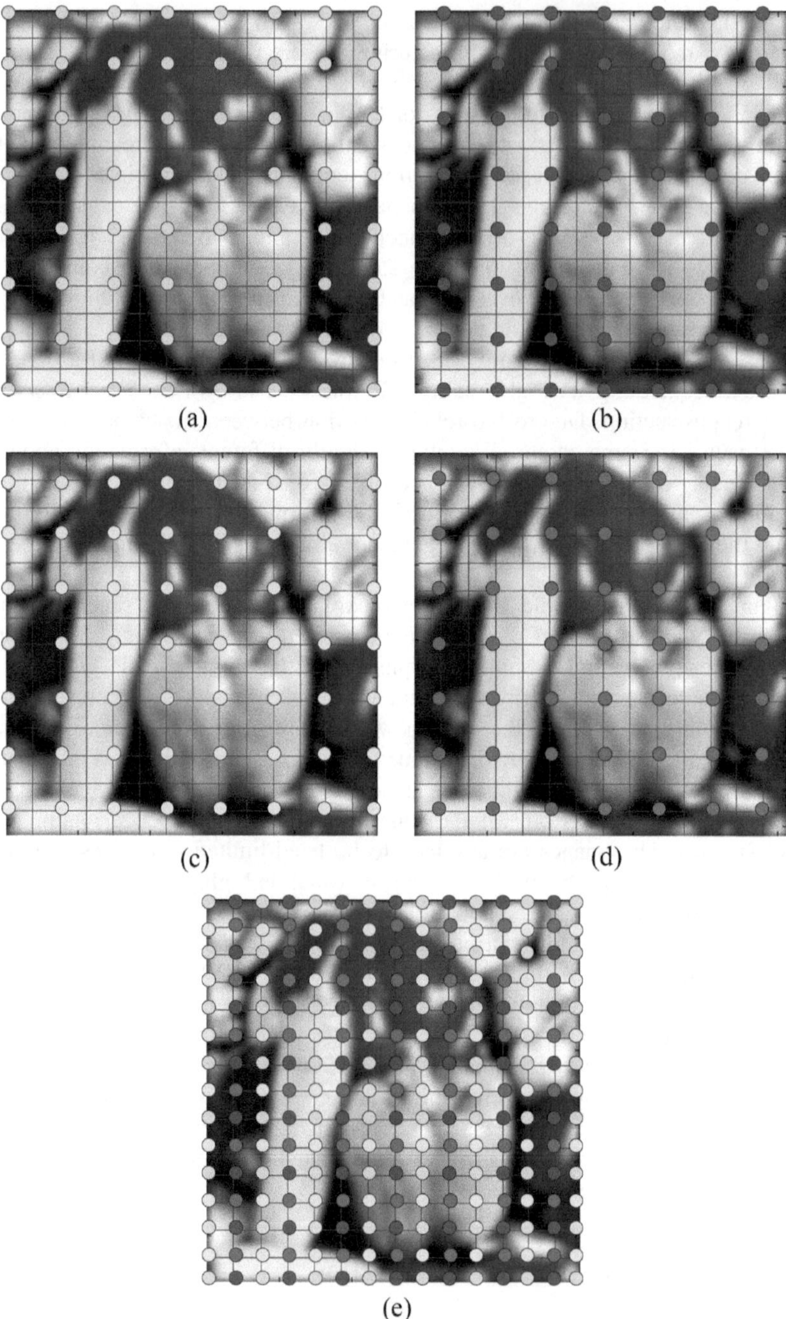

Fig. 1.5 A very simple illustration to explain the concept of super-resolution. (a)-to-(d) – Images acquired by shifting the camera. The coloured dots represent sample points. (e) By fusing the information contained within (a), (b), (c) and (d), we achieve a perfectly reconstructed high-resolution image.

computational time is of great importance and hence requires a super-resolution technique with high accuracy and low computational cost. On the other hand, for certain applications such as astronomical imaging or text recognition, computational time is not a constraint and therefore such applications can implement super-resolution techniques with high accuracy and a higher computational cost.

To name a few applications of super-resolution both in civilian and military domain:

- Medical Imaging.
- Remote Sensing.
- Target Detection and Recognition.
- Radar Imaging.
- Forensic Science.
- Surveillance Systems.

1.6 Research Outline

In this research, a novel approach to super-resolution image reconstruction is proposed. The problem of super-resolution is treated as an inverse problem, where we assume that LR frames are degraded versions of a HR image. In chapter 2, a review on various super-resolution techniques proposed in literature is presented and the research problem is defined.

In chapter 3, we present the theory for the proposed novel and hybrid reconstruction scheme, Iterative-Interpolation Super-Resolution, for the restoration of high-resolution images from a sequence of geometrically warped, aliased and under-sampled LR frames. Interpolation techniques are used to produce an initial estimate of the high-resolution image and then utilizing an iterative approach, the final approximation of the high-resolution image is produced. The reconstruction technique successfully handles the under-determined problem where only a relatively small number, as low as 10, of LR frames are available for accurate reconstruction for magnification factors as large as 20. Extensive analysis and high-resolution image results are presented to illustrate the success and accuracy of the proposed super-resolution image reconstruction scheme. Since simple matrix-vector operations are involved, the IISR system is highly computationally efficient making it suitable in programmable hardware for real-world applications.

In chapter 4, the problem of super-resolution is reformulated in terms of Tikhonov regularization optimization problem. The objective function is then minimized using an iterative conjugate gradient technique to produce the high-resolution imagery. Several forms of the regularization terms along with their effects on the reconstructed high-resolution imagery studied. Also, the full-solution of the proposed reconstruction scheme, described in chapter 3, is compared with the one produced using Tikhonov regularization to measure and evaluate the robustness and accuracy of reconstruction of the IISR scheme. Since optimization forms the basis for a majority of the super-resolution techniques, the comparison gives a general indication of performance of the IISR scheme over the optimization-based ones. Computer experiments and analytical data are presented.

In chapter 5, the problem of image registration is addressed which is a significantly critical pre-processing step in any super-resolution image reconstruction technique. Based on [5], affine motion model is considered and the motion parameters are estimated between the rotated and translationally shifted, aliased, under-sampled low-resolution frames. The registration technique is highly efficient due to its hierarchical architecture and the coarse-to-fine strategy. Computer experiments and analytical data are presented to illustrate the effectiveness and accuracy of the registration technique.

In chapter 6, based on the materials presented in this research, two professional softwares have been developed. The softwares were built using MATLAB, Image Processing Toolbox and TOMLAB (for optimization). The softwares were made completely modular to make them highly flexible. For a better explanation of the softwares, block diagrams and brief description of each module are also provided.

Finally, chapter 7 concludes this research and some ideas for future research are provided.

Chapter 2
Overview of Super-Resolution Techniques

2.1 Introduction

In the last two decades, extensive literature [6-10] has been produced on the research of super-resolution image reconstruction. The research has witnessed the introduction of many techniques in order to achieve high-resolution images from a sequence of geometrically warped, blurred, noisy, and under-sampled low-resolution images. Each of these techniques is either an extension of some previous methodologies or differs from the other techniques in their assumptions of the observation model or the type of reconstruction method applied for achieving the high-resolution images. On a broader aspect, we divide the various super-resolution techniques depending upon the domain (spatial or frequency) they are based on.

2.2 Frequency Domain – Based Approach to SR

Tsai and Huang [11] proposed the frequency domain approach for solving the problem of super-resolution image reconstruction in 1984. The frequency domain approach is based on an assumption that the original high-resolution image is band-limited and exploits the translational property of the Fourier Transform. It makes use of the aliasing relationship between the Continuous Fourier Transform (CFT) of the original real scene and the Discrete Fourier Transform (DFT) of the observed low-resolution images. In their paper, the sequence of low-resolution frames is assumed to be free from distortions such as blur or noise.

Let ***f(x, y)*** be a continuous high-resolution image and $f_k(x, y)$, where k = 1, 2, …, p, be a set of p translational shifted versions of ***f(x, y)***. Thus, considering arbitrary shifts δ_{xk} and δ_{yk} of ***f(x, y)*** along the x and y coordinates respectively [12],

$$f_k(x, y) = f(x + \delta_{xk}, y + \delta_{yk}) \quad \text{for } k = 1,2\ldots\text{p}. \tag{2.1}$$

The shifted image $f_k(x, y)$ was uniformly sampled using sampling periods T_1 and T_2 to generate the observed low resolution images,

$$f_k(i, j) = f(iT_1 + \delta_{xk}, jT_2 + \delta_{yk}). \tag{2.2}$$

V. Bannore: Iterative-Interpolation Super-Resolution Img. Recons., SCI 195, pp. 9 – 17.
springerlink.com

Using the shifting property of CFT, the CFT for $f_k(x, y)$, is given as:

$$F_k(u,v) = e^{j2\pi(\delta_{xk}u+\delta_{yk}v)} F(u,v). \tag{2.3}$$

Using the aliasing relationship and bandlimited constraint,

$$|F(u,v)| = 0, \; for \; |u| \geq L_x w_x, |v| \geq L_y w_y, \tag{2.4}$$

the relationship between the CFT of the high-resolution image and the DFT of the kth observed low-resolution image can be expressed as,

$$F_k(m,n) = \frac{1}{T_1T_2} \sum_{i=-L_x}^{L_x-1} \sum_{l=-L_y}^{L_y-1} F_k\left(\frac{2\pi m}{MT_1} + iw_x, \frac{2\pi n}{NT_2} + lw_y\right) \text{ for } k = 1,2\ldots p, \tag{2.5}$$

where $w_x = 2\pi / T_1$ and $w_y = 2\pi / T_2$. By using lexicographic ordering of (2.5), we get,

$$b = AX, \tag{2.6}$$

where, $\boldsymbol{b}$ is a column vector with the kth element of the DFT coefficients of $\boldsymbol{F_k(m,n)}$, $\boldsymbol{X}$ is a column vector with the samples of the unknown CFT of $f(x, y)$ and $\boldsymbol{A}$ is a matrix which relates the DFT of observed low-resolution images to the samples of the continuous high-resolution image. Thus, the reconstruction of the high-resolution image $\boldsymbol{X}$ requires calculating the DFT's of the observed low-resolution images, estimate the matrix $\boldsymbol{A}$ and solve eq. (2.6) which is an inverse problem.

An extension of [11] was proposed by Kim *et al.* [12] in which they introduced a weighted recursive least square algorithm based on the aliasing relationship between the low-resolution images and high-resolution image, for reconstructing Super-Resolution image from available noisy under-sampled images. The algorithm combines filtering and reconstruction. The algorithm considered only noise and not blur. The idea was further developed by Kim and Su [13] where they considered blur and noise together as distortions in the observed low-resolution images. Also, to stabilize the ill-conditioned nature of the inverse problem of super-resolution, the recursive algorithm of [12] was refined to include iterative update of the regularization term.

Frequency – based techniques for estimating high-resolution images are simple to implement and directly addresses removal of aliasing artifacts. They have very low computational complexity, thereby making them highly capable of parallel implementation. Although the technique has the applicability to real-time applications, it is confined to only global translational motion and has limited ability to apply spatial domain *a priori* knowledge in the regularization term.

2.3 Spatial Domain – Based Approach to SR

Spatial domain techniques are the most popular ones developed for super-resolution. The popularity is due to the fact that the motion is not limited to translational shifts only and thus a more general, global or non-global motion can also be incorporated and dealt with.

2.3.1 *The Model*

The classic model of super-resolution in the spatial domain assumes that a sequence of N low-resolution images represent different snapshots of the same scene. The real scene to be estimated is represented by a single high-resolution reference image $\underline{\boldsymbol{X}}$ of size (P by P). Each LR frame, $b_1, b_2, \ldots b_N$, is a noisy, down-sampled version of the reference image that is subjected to various imaging conditions such as optical, sensor and atmospheric blur, motion effects, and geometric warping. The size of each LR frame is (M by M) and M < P. It is convenient to represent the observation model in matrix notation:

$$\begin{bmatrix} b_1 \\ \vdots \\ b_N \end{bmatrix} = \begin{bmatrix} D_1 \cdot B_1 \cdot W_1 \\ \vdots \\ D_N \cdot B_N \cdot W_N \end{bmatrix} \underline{X} + \begin{bmatrix} e_1 \\ \vdots \\ e_N \end{bmatrix} = \begin{bmatrix} A_1 \\ \vdots \\ A_N \end{bmatrix} \underline{X} + \begin{bmatrix} e_1 \\ \vdots \\ e_N \end{bmatrix}, \tag{2.7}$$

where, $\boldsymbol{D}$ represents the down-sampling matrix of size (M^2 by P^2), $\boldsymbol{B}$ is the blur matrix of size (P^2 by P^2) and $\boldsymbol{W}$ is the matrix representing geometric warping of size (P^2 by P^2). By grouping and rewriting eq. (2.7), the model is given as:

$$\underline{b} = A\underline{X} + \underline{e}, \quad A_k = \left[D_k \cdot B_k \cdot W_k \right]_{for\ k=1,2\ldots N}. \tag{2.8}$$

In eq. (2.8), linear operator $\boldsymbol{A}$ (size - M^2 by P^2), represents sub-sampling, motion compensation and all the other imaging factors, the LR frames are given by $\underline{\boldsymbol{b}}$ (size - M^2 by 1) and the additive Gaussian noise is represented by $\underline{\boldsymbol{e}}$ (size - M^2 by 1). The images are represented in eq. (2.8) as vectors, shown by an underscore, and are ordered column-wise lexicographically.

2.3.2 *Iterative Back-Projection Techniques*

Super-resolution of monochrome and colour low resolution image sequences was considered by Irani and Peleg [14]. They derived an iterative back-projection algorithm based on computer aided tomography. The algorithm starts with an initial guess ($\boldsymbol{X^0}$) for the output high-resolution image and the imaging process ($\boldsymbol{A}$) is simulated to generate low-resolution images ($\boldsymbol{b^{sim}}$) based on the initial guess. These simulated low-resolution images are then compared with the observed ones ($\boldsymbol{b}$) and the error generated between them is back-projected onto the initial guess via back-projection operator ($\boldsymbol{A^{bp}}$), thereby minimizing the error iteratively.

$$X^1 = X^0 + A^{bp}(b - b^{sim}). \tag{2.9}$$

The algorithm considers translational and rotational motion but the authors claim that the same concept can be applied to other motions also. They considered multiple motion analysis in [15] including occlusion and transparency. The algorithm successfully solves the issue of blur and noise, however due to the ill-posed nature of super-resolution, the technique is unable to generate a unique solution.

2.3.3 Optimization Techniques

The problem of estimating a high-resolution image from a sequence of low-resolution images is an inverse problem which is highly ill-conditioned. To stabilize this ill-posed nature, *Regularization* term is included. The problem of super-resolution, shown in eq. (2.8), is then solved, in the least-squares sense, by minimization of the error τ along with the addition of the regularization term. The modified equation of (2.8) is given as:

$$\tau = \sum_{k=1}^{N} [\underline{b}_k - A_k \underline{X}]^2 + \lambda [Q \underline{X}]^2 . \tag{2.10}$$

In the above equation, $\boldsymbol{Q}$ is the regularization or stabilization matrix and $\lambda > 0$ is the regularization parameter. Although there is no unique procedure for constructing the regularization term, it is usually chosen to incorporate *a priori* knowledge of the real high-resolution scene, such as the degree of smoothness.

In the case of Forward Looking Infrared (FLIR), images are spatially under-sampled and are degraded due to aliasing effect. Alam *et al.* [6] used weighted-nearest-neighbor and Weiner filter for estimating high-resolution images from FLIR images. An extension of this algorithm was proposed in [16] which considered a more general motion, translational and rotational. To refine it further, they introduced a simple regularization term into their cost function that enforced the smoothness of the final solution. They used steepest descent and conjugate-gradient techniques for minimization of the error between the simulated low-resolution images and the observed ones. Alam *et al.* [17] proposed a more efficient real time applicable registration and high-resolution reconstruction technique using multiple random translational shifted frames. They used weighted-nearest-neighbor and Weiner filter for estimating the high-resolution images.

2.3.4 Bayesian Techniques

The Bayesian techniques are based on the Bayes' Theorem and treat the problem of estimating the high-resolution image as a statistical estimation problem. These techniques provide a convenient way to include *a priori* knowledge as constraints to stabilize the ill-posed nature of super-resolution image reconstruction.

The classic model of super-resolution defined in eq. (2.8) can be rewritten as:

$$b = AX + e . \tag{2.11}$$

The *Maximum A-Posteriori* (MAP) estimator for the high-resolution image maximizes the *a-posteriori* probability $\mathbf{P\{X \mid b\}}$,

$$\hat{X}_{MAP} = \arg\max_X P\{X \mid b\}. \tag{2.12}$$

Now, applying the Bayes' theorem to eq. (2.12),

$$\hat{X}_{MAP} = \arg\max_X \left[\frac{P\{b \mid X\}P\{X\}}{P\{b\}}\right]. \tag{2.13}$$

Taking the logarithmic function,

$$\hat{X}_{MAP} = \arg\max_X \left[\log P\{b \mid X\} + \log P\{X\}\right]. \tag{2.14}$$

In eq. (2.14), on the right-hand side of the equation, the term **log P{X | b}**, is known as the log likelihood function and the term **log P{X}** is referred to as the log of the *a priori* image model. The MAP estimation model provides the ability to include *a priori* knowledge and thus, effectively regularizes the ill-conditioned nature of super-resolution reconstruction.

A MAP framework for estimating the high-resolution scene was proposed in [18, 19]. In [18], the regularization parameter was fixed and gradient descent was the method chosen for minimization purpose. However, in [19], an iterative update of the regularization parameter was proposed which does not depend upon any *a priori* information but updates itself with the enhanced image generated from their iterative gradient reconstruction process.

Another Bayesian technique used in super-resolution reconstruction is the *Maximum-Likelihood* (ML) technique. ML is a special extension of the MAP framework without the priori term. In [20, 21], ML technique was proposed for the estimation of high-resolution images solved by Expectation-Maximization (EM) algorithm. In [20], they combined the process of estimating the sub-pixel shifts between the low-resolution images and restoring the images from degradations such as noise and blur. Later, in [21], the above two processes were combined with the reconstruction of the high-resolution image. The Bayesian techniques offer a robust environment for modelling noise and also provide the ability to include *a priori* knowledge thereby stabilizing the ill-conditioned nature of super-resolution. Due to the advantage of inclusion of prior information, MAP framework is usually preferred over ML.

2.3.5 Projection onto Convex Set (POCS) Technique

The method of Projection onto Convex Sets, popularly known as POCS was introduced by [22, 23] in 1982. In [24, 25], Stark explains the general technique for applying POCS in the field of image restoration. The concept of POCS applied to the problem of super-resolution was first introduced in [26]. According to the method of POCS for super-resolution reconstruction, the space of estimated high-resolution solutions is restricted by a set of constraints (closed convex sets) which characterize desirable properties, such as fidelity to data, smoothness, sharpness etc., to be consistent in the final solution. For each set of convex constraints $\boldsymbol{C_i}$, a projection operator $\boldsymbol{T_i}$ is defined. The problem is then reduced to iteratively locate, given a point in the high-resolution image space, the closest solution which intersects with all the given convex constraints, $\boldsymbol{C_i}$. The convergence can be given as:

$$
\begin{aligned}
& X^{n+1} = T_i X^n \\
& \Rightarrow X^{n+1} = T_m T_{m-1} \ldots . T_2 T_1 X^n \quad \textit{for } n = 0,1,2.....
\end{aligned}
\tag{2.15}
$$

Recently POCS has been used in improving resolution from multi-camera surveillance imaging [27]. The method of POCS is simple and allows convenient inclusion of *a priori* information but has a very high computational cost and a slow convergence rate which limits its practical applicability. Also, the final solution is not unique and highly depends upon the initial guess.

2.3.6 Preconditioned Techniques in Optimization

The problem of super-resolution is an inverse problem and the matrix $\boldsymbol{A}$ (eq. 2.8) contains very small singular values which are either zero (due to round off error) or approaching zero. This characteristic of matrix $\boldsymbol{A}$ makes it singular in nature and highly ill-conditioned. Due to this, the solution is very sensitive and can vary tremendously in an arbitrary manner with very small changes in the data. Thus, to make the solution unique and stable, another term is added to eq. (2.8) known as the Regularization term as shown in eq. (2.10). Most of the inverse problems (such as super-resolution) are ill-posed and the solution is tremendously sensitive to the data. This also leads to the problem of slow convergence and high computational load.

In general, preconditioners are approximate inverses which convert the ill-conditioned matrices, such as matrix $\boldsymbol{A}$, to a well-behaved one. Introducing preconditioners in an ill-posed problem removes the need for employing regularization techniques for stabilizing the solution. Essentially, by multiplying a preconditioner, say $\boldsymbol{M}$ with matrix $\boldsymbol{A}$, the condition number of this product, $\boldsymbol{M^{-1}A}$ is smaller than that of matrix $\boldsymbol{A}$ alone. Such preconditioner not only makes an ill-conditioned system of equations to a well-behaved one but also increases the rate of convergence and decreases the computational load.

In [28, 29], preconditioning of conjugate gradient using efficient block-circulant preconditioners was proposed. As per the authors, preconditioning is a technique used to transform the original system into one with the same solution, which can be solved by the iterative solver more efficiently and quickly. The unconditioned system starting from ground zero, takes a longer time to converge (due to the high computational cost involved) as compared to a preconditioned system. FFT is employed for computing the preconditioners which thereby reduces the overall computational time of the system. Preconditioning certainly provides a promising approach for the use of super-resolution in real-time environment.

2.3.7 *Other SR Techniques*

2.3.7.1 Hybrid (ML + POCS) Technique

In [7], a hybrid algorithm using ML and POCS was proposed for estimating a high-resolution image from multiple blurred, noisy, under-sampled low-resolution images. The hybrid technique combined the benefits of the Bayesian and the set theoretic technique. With the help of POCS, all the *a priori* knowledge could be utilized beneficially and a single optimal solution could be reached.

2.3.7.2 Adaptive Filtering Technique

Adaptive filtering approach applied to the time axis for super-resolution reconstruction was introduced in [30]. The authors proposed a few algorithms based on least squares - recursive least squares (RLS) and pseudo-RLS. For estimating the high-resolution image both steepest descent (SD) and normalized steepest descent (NSD) were applied. According to the authors, their approach allows treating linear time and space variant blurring and general motion. In their further research, the authors used Kalman filtering approach [31, 32] for solving the problem of super-resolution. In [32] they re-derived R-SD and R-LMS as approximations of the Kalman filter. The algorithms were built with the assumption that the information regarding the motion between the images and the blur operators is known which otherwise in reality would need to be estimated to use these algorithms. This assumption is very critical for the performance of the algorithms and significantly depends on it. The Kalman filter approach is promising but is still in an experimental state as its computational cost is extremely high.

2.3.7.3 Learning-Based Techniques

Even though a major class of super-resolution techniques is reconstruction-based, in recent years, a lot of interest has grown towards learning-based techniques. This class of super-resolution technique was initially introduced in [33-35]. This technique generates high-resolution images from one or more low-resolution frames by learning from a collection of training images or strong image priors learned from data before. These training images are scenes of either the same or different types. Freeman *et al.* proposed a learning-based super-resolution technique, known as Vision by Image/Scene TrAining (VISTA), in [33] for low-level vision problems. The learning scheme is based on the Markov network. In [35], Freeman *et al.* proposed an example-based learning technique which used the Markov network to model the relationships between the high- and low- resolution images, and between neighbouring high-resolution patches. In [34], the authors proposed a learning-based super-resolution technique for human faces or text and termed it as face hallucination or recogstruction. Using a face database, the technique learns the corresponding relationship between the low-resolution frames of human faces and their known high-resolution frames. It is this learned information which is later utilized for the reconstruction of high-resolution images from low-resolution images of human faces.

Inspired by the recent manifold learning methods, Chang *et al.* proposed a single-frame super-resolution technique in [36], based on Locally Linear Embedding (LLE). The technique assumes that the image patches in both the low- and high- resolution frames form manifolds with similar geometry in two distinct spaces. The technique allows the usage of multiple training samples to contribute simultaneously for the generation of each image patch in the high-resolution image. Due to this unique quality, the technique proposed in [36] requires fewer training samples as compared to other learning-based techniques. In [37], Qiao *et al.* proposed a blind single-frame super-resolution technique which is combined with shadow removal in a single operation. To eliminate the lighting effects of the image, the authors utilize logarithmic-wavelet transform for constructing a manifold structure. The technique is claimed to be robust to quantization errors and less sensitive to the training sets. The performance of learning-based super-resolution techniques depends on the accuracy of matching between the input low-resolution frame and the training samples. Despite this fact, the popularity of learning-based techniques is bound to increase since they can successfully achieve super-resolution utilizing as low as one low-resolution frame.

2.4 Research Problem

An imaging system is only able to capture a true or natural scene with only finite levels of resolution as compared to the true or natural scene which has infinite levels of resolution. There are two solutions that can be implemented to increase the resolution of an imaging system - hardware or software. The best solution is the hardware improvement of the imaging system to achieve higher resolution. However, it is not always feasible to achieve higher resolution by hardware enhancement due to practical reasons. Therefore, in such a scenario, it is feasible to employ an intelligent software solution to generate higher resolution than what is captured by the imaging system. One such software solution is super-resolution image reconstruction.

Super-resolution image reconstruction refers to image processing techniques that attempts to reconstruct high quality, high-resolution images by utilising incomplete scene information contained in a sequence of geometrically warped, aliased, and under-sampled low-resolution images. This estimation of high-resolution image is also referred to as an inverse problem.

A common feature of such inverse problems is their sensitivity to even small perturbations of the data that may introduce significant errors in the reconstruction process. This makes the process ill-conditioned and intrinsically unstable. Because of this feature, most of the research on super-resolution has been directed towards increasing the robustness and the fidelity of the reconstruction process. Much less attention has been devoted to more practical issues of computational efficiency and real-time applicability of super-resolution. Yet, super-resolution reconstruction is a very computationally intensive process that has to deal with big data sets and inherent instabilities. As a result of this, many of the developed techniques are not suitable for practical applications that require real-time or even reasonably fast processing. It is, thus, desirable to develop algorithms that maintain a proper balance between computational performance and the fidelity of the reconstruction.

Therefore in this study, the motivation for research is to explore possible new approaches for a computationally efficient yet fairly accurate super-resolution image reconstruction process to generate a high-resolution image. Also, super-resolution being an inverse and ill-conditioned problem, the optimization based technique for super-resolution is reinvestigated and the significant influence of the regularization term over the fidelity of reconstruction is studied.

Image registration is a critical pre-processing step in image super-resolution. Even though, it is not the primary aim of this research, we address the issue for estimating the relative motion parameters between rotated and translationally shifted low-resolution images.

Chapter 3
Iterative-Interpolation Super-Resolution (IISR)

3.1 Image Interpolation

In the Oxford dictionary, the word interpolation has the meaning - "*to insert (an intermediate term) into a series by estimating or calculating it from surrounding known values*". In the digital age, image interpolation refers to the technique of recovering a continuous signal by estimating image data from a set of discrete image data samples. It links the continuous and the discrete domains. Image interpolation forms a fundamental base in image processing and is the heart of many computer vision applications such as medical imaging, target detection and recognition, and astronomical imaging. Almost every image processing software implements some interpolation technique for transformations, rotations and many other manipulations performed on an image. It is very important that the interpolation techniques have a very low computational cost in terms of both, time and memory utilization since they are usually implemented at some intermediate step in a system. At the same time, it is necessary for the technique to yield good and accurate results, or else it could jeopardize the final solution. For example, in the field of medical imaging, computed tomography (CT) or computed axial tomography (CAT) and magnetic resonance imaging (MRI) scan employ interpolation techniques during the registration process, a slight error in the interpolated data could cause mis-registration thereby significantly affecting the accuracy of reconstruction of the final image which may lead to wrong diagnosis of a patient. It is therefore very important to choose a correct type of interpolation technique, depending upon the nature of its application, which provides the best trade-off between accuracy and computational cost.

As per the Shannon-Nyquist sampling theorem [2-4, 38, 39], a continuous signal (band-limited) can be completely recovered from its samples, if the sampling frequency is twice the highest frequency (Nyquist frequency) in the original signal. For a 1D case, let $f(x)$ be the continuous signal to be reconstructed from its samples $f_k(m)$, where k = 1, 2, …, p. The interpolation process in terms of convolution in the spatial domain can then be given as,

$$f(x) = f_k(m) * h(x) \quad \text{for } k = 1,2\ldots\text{p}. \tag{3.1}$$

where $h(x)$ is the interpolation or reconstruction kernel.

V. Bannore: Iterative-Interpolation Super-Resolution Img. Recons., SCI 195, pp. 19 – 50.
springerlink.com

In order for the interpolation technique to reconstruct the continuous signal from its discrete data samples, the kernel should be,

- Symmetric

$$h(x) = h(-x). \tag{3.2}$$

- Zero for all non-zero integers and one if its argument is zero. This rule ensures that the interpolation coefficients become the sampled data points.

$$h(x) = \begin{cases} 1, & if\ x = 0 \\ 0, & if\ x \neq 0 \end{cases}. \tag{3.3}$$

- Separable in order to reduce computational cost. For example, a 2D interpolation kernel is given by ***h(x,y)***, then they are separable as:

$$h(x, y) = h(x) \bullet h(y). \tag{3.4}$$

Therefore, using eq. (3.4) for 2D, eq. (3.1) can be rewritten as:

$$\begin{aligned} f(x, y) &= f_k(m,n) * h(x, y) \quad \rightarrow k = 1, 2....p \\ &= (\ (f_k(m,n) *_x h(x)) *_y h(y)\ ) \end{aligned}, \tag{3.5}$$

where $*_x\ and\ *_y$ denotes convolution in x and y direction respectively.

It is a fact that the type, size and shape of the kernel chosen for interpolation are major factors contributing to the reconstruction quality of the final image or signal. Generally, the size of the interpolation kernel is very crucial since it determines the computational cost of the system.

3.2 Interpolation Convolution Kernels

In this section we describe some of the most well-known interpolation techniques that can be utilized as convolution kernels.

3.2.1 Linear Interpolation

This is one of the most popular interpolation techniques. Due to the low-level complexity of this technique it has gained a lot of popularity. The general expression is given as:

$$f(x) = \begin{cases} 1 - |x| & \rightarrow 0 \leq |x| < 1 \\ 0 & \rightarrow elsewhere \end{cases}. \tag{3.6}$$

In 2D, this interpolation technique is known as *Bilinear.* This technique cuts-off all the high-frequencies. The frequency response for linear interpolation kernel is shown in fig. 3.1(e).

3.2.2 Nearest - Neighbor Interpolation

Nearest-neighbor technique simply uses the value from the nearest pixel. In terms of a convolution kernel this is a rectangular function, with width of one pixel. The general expression is given as

$$f(x)=\begin{cases}1 & \rightarrow -0.5\leq|x|<0.5\\ 0 & \rightarrow elsewhere\end{cases}. \tag{3.7}$$

This technique causes aliasing and blurring effects in the interpolated image. The technique is very simple and easy to implement but at the cost of severe loss of quality. The frequency response for nearest-neighbor interpolation kernel is shown in fig. 3.1(f).

3.2.3 Sinc Interpolation

This technique uses the *Sinc* function. The name *Sinc* comes from "Sine Cardinal". The normalized *Sinc* function is written as:

$$Sinc(x)=\begin{cases}1 & \rightarrow x=0\\ Sin(\pi x)/(\pi x) & \rightarrow x\neq 0\end{cases}. \tag{3.8}$$

The Sinc function is also known as the ideal reconstruction filter. The function is symmetric, Sinc(x) = Sinc (-x). Also, the Sinc function is zero for all integer values of its argument except for zero. Since the Sinc filter is spatially unlimited and has an infinite impulse response, the function is impracticable. To solve this problem, the Sinc function can be multiplied by a function which is non-zero in a finite range. This function is referred to as the Window Function and the Sinc is then known as Windowed Sinc Function.

In this research, Lanczos window of degree 2 and 3 are utilized. Lanczos is the central lobe of the scaled Sinc function. It is given by:

$$f(x)=\begin{cases}Sin(\pi x/d)/(\pi x/d) & \rightarrow |x|<d\\ 0 & \rightarrow else\end{cases}, \tag{3.9}$$

where, ***d*** is the degree of Lanczos window. The frequency responses for Lanczos window interpolation kernel (degree 2 and 3) are shown in fig. 3.1(a-b), respectively.

3.2.4 Cubic-Spline Interpolation

The cubic interpolation technique is composed of the piecewise cubic polynomials defined on (-2,-1), (-1, 0), (0, 1) and (1, 2). Beyond (-2, 2), the output of the cubic interpolation kernel is zero. The cubic interpolation is sometimes also referred to as the *Keys* function. The general expression for this interpolation technique is,

$$f(x)=\begin{cases}(a+2)|x|^3-(a+3)|x|^2+1 & \rightarrow 0<|x|<1 \\ (a)|x|^3-5a|x|^2+8a|x|-4a & \rightarrow 1<|x|<2 \\ 0 & \rightarrow elsewhere\end{cases}. \tag{3.10}$$

The above expression depends upon the choice of "***a***". The best choice for '***a***' is "*-0.5*", which forms the third-degree cubic kernel [40]. The kernel is zero for all integer values of its argument and is one at zero. The kernel has finite support and is symmetric. This cubic kernel also attracts more attention because of its strictly positive nature [41]. In various applications such as image processing, the pixels carry the image intensities which should always be non-negative. The cubic kernel always being strictly positive ensures that the interpolated image is positive. The frequency responses for cubic-spline interpolation kernel (order 4 and 6) are shown in fig. 3.1(c-d), respectively.

3.2.5 Gaussian Interpolation

In 1996, a new approach was developed for the generation of interpolation kernels [42]. It exploited the *Gaussian* function in both the signal and the Fourier space. As per the author, the interpolation kernel has to satisfy 3 main goals:

- Locally compact in signal space.
- Locally compact in Fourier space.
- Easy mathematical manipulation.

A unit area Gaussian Kernel with zero mean and variance β is given by (3.11),

$$G_0(x,\beta)=\frac{1}{\sqrt{2\pi\beta}}e^{-x^2/2\beta}. \tag{3.11}$$

The successive derivatives are:

$$G_m(x,\beta)=\frac{\partial^m}{\partial x^m}G_0(x,\beta). \tag{3.12}$$

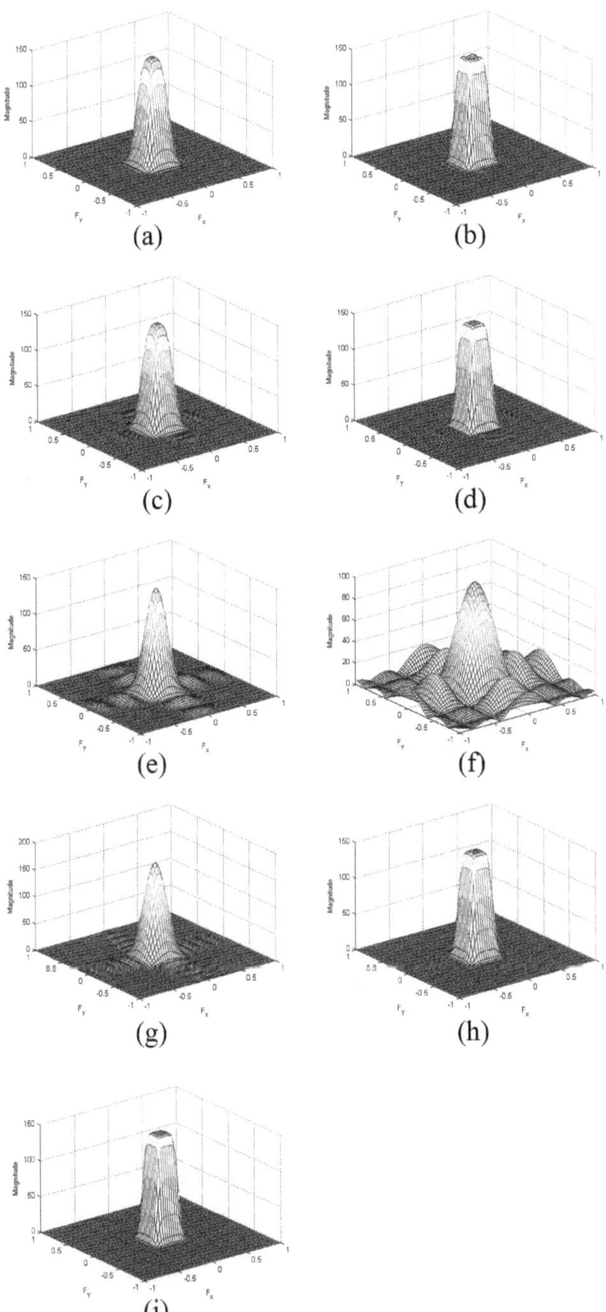

Fig. 3.1 Frequency response of interpolation kernels - (a) Sinc Lanczos: degree 2, (b) Sinc Lanczos: degree 3, (c) Cubic Polynomial: order 4, (d) Cubic Polynomial: order 6, (e) Linear, (f) Nearest Neighbor, (g) Gaussian: order 2, (h) Gaussian: order 6 and (i) Gaussian: order 10.

The expressions for the 2nd, 6th and 10th order Gaussian kernel are given by:

$$^{Gauss}f_2(x) = G_0(x, 2\gamma_2) - \gamma_2 G_2(x, \gamma_2)$$

$$^{Gauss}f_6(x) = G_0(x, 2\gamma_6) - \gamma_6 G_2(x, \gamma_6) - \frac{\gamma_6^{\,3}}{24} G_6(x, \gamma_6) \tag{3.13}$$

$$^{Gauss}f_{10}(x) = G_0(x, 2\gamma_{10}) - \gamma_{10} G_2(x, \gamma_{10}) - \frac{\gamma_{10}^{\,3}}{24} G_6(x, \gamma_{10}) - \frac{\gamma_6^{\,5}}{1920} G_{10}(x, \gamma_{10})$$

where $\gamma_2 \approx 0.464, \gamma_6 \approx 0.866 \;\&\; \gamma_{10} \approx 1.269$. The frequency responses for the Gaussian interpolation kernel (order 2, 6, and 10) are shown in fig. 3.1(g-h-i), respectively.

3.3 The Imaging Model

The observation model for super-resolution reconstruction simulates the physical process of image acquisition. It is assumed that a sequence of N low-resolution images is captured by a camera moving over a scene of interest. The objective is to reconstruct a high-resolution representation of the original scene from degraded and incomplete data represented by the LR frames. The original scene is represented by a HR reference image ***X*** of size (P by P) that we want to reconstruct. Each LR image, $b_1, b_2, \ldots, b_N$, is the result of sampling (decimation) of the HR reference image and is subjected to various degrading factors such as optical, sensor and atmospheric blur, motion effects, and geometric warping. The size of each LR frame is (M by M) and $M < P$. It is convenient to represent the observation model in matrix notation:

$$\begin{bmatrix} b_1 \\ \vdots \\ b_N \end{bmatrix} = \begin{bmatrix} D_1 \cdot B_1 \cdot W_1 \\ \vdots \\ D_N \cdot B_N \cdot W_N \end{bmatrix} \underline{X} + \begin{bmatrix} e_1 \\ \vdots \\ e_N \end{bmatrix} = \begin{bmatrix} A_1 \\ \vdots \\ A_N \end{bmatrix} \underline{X} + \begin{bmatrix} e_1 \\ \vdots \\ e_N \end{bmatrix}, \tag{3.14}$$

where, ***D*** represents the down-sampling matrix of size (M^2 by P^2), ***B*** is the blur matrix of size (P^2 by P^2) and ***W*** is the matrix representing geometric warping of size (P^2 by P^2). If the resulting set of N low-resolution images is abbreviated by $\underline{\boldsymbol{b}} = [b_1, b_2, \ldots\ b_N]^T$ (size - M^2 by 1), and that the imaging process (sub-sampling, motion compensation and all the other imaging factors) is represented by a linear operator, ***A*** of size (M^2 by P^2), then the mathematical representation of the super-resolution model can be rewritten as,

$$\underline{b} = A\underline{X} + \underline{e}, \qquad A_k = \left[D_k \cdot B_k \cdot W_k \right]_{for\ k=1,2...N}, \tag{3.15}$$

where, $\underline{e}$ (size - M^2 by 1) represents the additive Gaussian noise. In eq. (3.15), images are represented as vectors by an underscore and are ordered column-wise lexicographically.

3.4 Interpolation-Based Super-Resolution

Super-resolution image reconstruction process consists of several stages. At the initial stage, the camera motion between the scene and the camera sensor has to be taken into account.

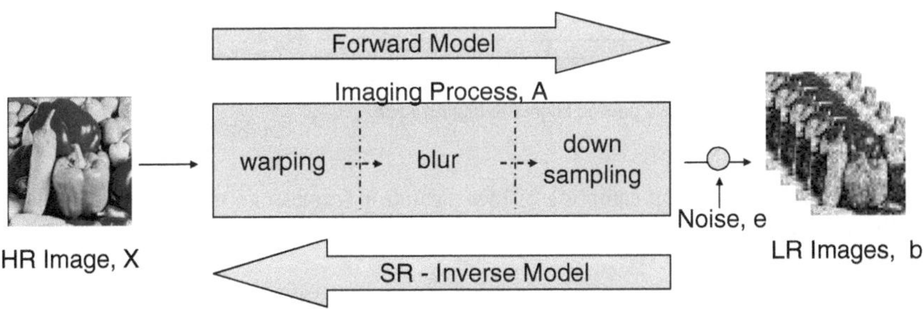

Fig. 3.2 Super-resolution (SR) Imaging Model

To accomplish this, the sequence of geometrically warped, under-sampled, low-resolution frames is registered precisely in reference to a low-resolution frame. Once registered, each pixel from each of the low-resolution frames is then placed onto a high-resolution composite grid using the registration information estimated by the registration routines. This step is illustrated in fig. 3.3, where for simplicity it is assumed that the relative shifts between LR frames are smaller than the up-sampling factor. It is important to mention here, that, accurate and precise registration affects the quality of high-resolution image reconstruction significantly, since a mis-registration will cause the pixels to be placed incorrectly in the composite grid which will not only cause loss of information but will also introduce errors during the reconstruction process leading to an inaccurate solution. Image registration is discussed in detail in chapter 5.

The randomness of sampling due to camera movements is the core of super-resolution. In order to address the randomness, the research aims at the underdetermined problem of super-resolution where the number of LR frames is insufficient to fill up all HR pixels. If the total number of LR pixels (in all LR frames) were sufficient to fill up all the high-resolution pixels, in the presence of

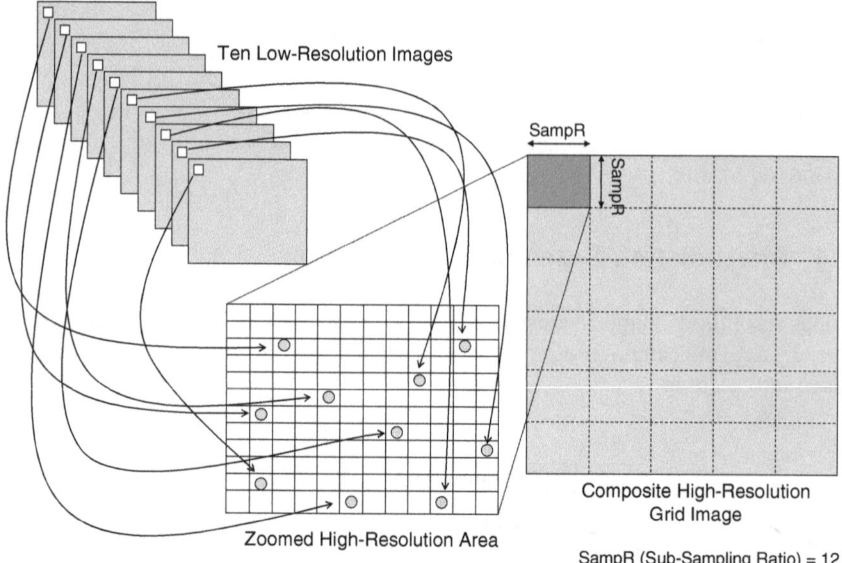

Fig. 3.3 Translation and up-sampling of low-resolution frames into the high-resolution grid for the up-sampling ratio =12

blur and noise, the reconstruction would be trivial and the problem of super-resolution would then be reduced to the classical restoration of images by debluring and noise reduction. It is therefore assumed that the number of LR frames is limited and are insufficient to fill up the entire HR composite grid. The *first approximation*, X_1, of the high-resolution image is then estimated by interpolating the sparse composite grid (convolution between the composite grid image and the interpolation kernel) to populate the empty pixels. This is an important step that has considerable effect on the accuracy of the reconstruction. The quality of the high-resolution image is directly proportional to the number of LR frames used in the reconstruction process; higher the number of LR frames, better is the reconstruction quality.

It is worth mentioning at this point that the resolution enhancement of the LR images achieved in the multiframe reconstruction is usually smaller than the up-sampling ratio and is limited by a number of the LR frames (which we assume to be not too large). As a rule of thumb, the linear resolution enhancement is proportional to $\sqrt{N}$. The size of the interpolating kernel has to be congruent with N to recover spatial frequencies up to the expected Nyquist frequency of the reconstructed image (which is smaller than the Nyquist frequency of the HR grid) and suppress higher frequencies. For practical reasons, however, it is important to keep the number of low-resolution images as low as possible because with large number of LR frames, the accumulation of errors would impede the reconstruction accuracy.

In order to quantify the accuracy of the reconstruction, a synthetic set of low-resolution images was generated for each high-resolution image by applying random translations. The reconstruction procedure was then applied to the low-resolution images and the reconstructed high-resolution image was compared with the original. The interpolation techniques described in subsection 3.2 were implemented. Figure 3.4 shows one of the low-resolution image frames used for reconstruction of a high-resolution image. In this particular example 10 LR frames were used for the reconstruction of the high-resolution image and the decimation ratio was 12. Figure 3.5 shows the HR

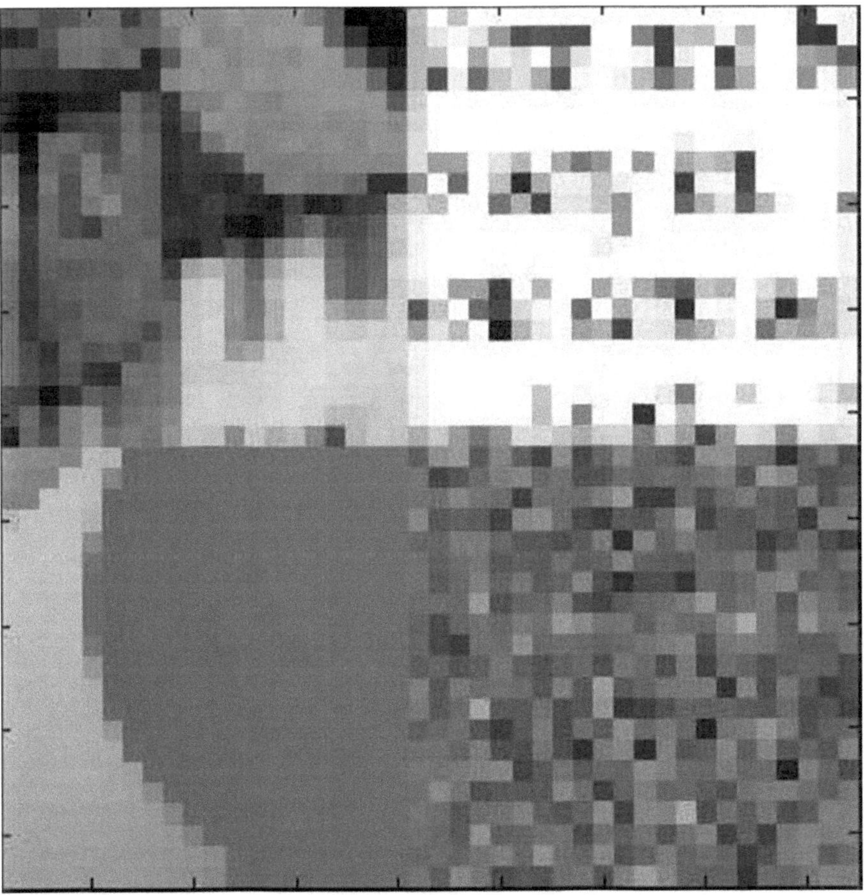

Fig. 3.4 One of the 10 low-resolution frames generated with sampling ratio, 12

composite grid image created by mapping all the pixels from all the LR frames using the accurate registration information.

The initial reconstructed HR image, X_l, for fig. 3.4 panels is shown in fig. 3.6. By inspecting the reconstructed images one can clearly notice the deficiency of the reconstruction process. The periodic artifacts that are visible in the interpolated images are due to the high number of empty blocks in the composite high-resolution grid (see fig. 3.5). These artifacts repeat after a period of 12 pixels (decimation or down-sampling ratio) and are the result of the irregular sampling of the scene due to the simulated random movements of the camera. The interpolation process, in this case, is unable to efficiently cope up with the randomness and the sparsity of data in the predominantly empty HR composite grid. Therefore, we adopt an iterative approach to image super-resolution.

Fig. 3.5 High-resolution composite grid image

Fig. 3.6 First approximation interpolated image

3.5 Iterative-Interpolation Super-Resolution

A solution of the least-square problem can be sought out by solving the equivalent normal equation, which involves the inversion of a linear operator. Consider that only an approximate inverse is known, the solution can then be improved by an iterative approach described in [43]. In the context of super-resolution reconstruction, the approximate inverse operation is given by our interpolation-based reconstruction, as described in the previous section. The iterative approach technique also tackles the problem of visible artifacts and improves the quality of the high-resolution image.

Ignoring the noise term, $\boldsymbol{e}$ of eq. (3.15) and let $\boldsymbol{X}$ be the exact solution,

$$A \cdot X = b \,, \tag{3.16}$$

where, $\boldsymbol{A}$ is a matrix and $\boldsymbol{X'}$ is an estimated inaccurate solution of $\boldsymbol{X}$, such that $\mathbf{X} = X' + \delta\boldsymbol{X}$. Substituting the value of $\boldsymbol{X}$ in eq. (3.16), we get

$$\begin{aligned} A\cdot(X' + \delta X) &= b \\ \Rightarrow \quad A\cdot\delta X &= b - A\cdot X' \end{aligned} \qquad (3.17)$$

Eq. (3.17) can now be solved for δX, as an iterative approach,

$$\begin{aligned} A\cdot(X_{k+1} - X_k) &= b - A\cdot X_k \\ \Rightarrow \quad (X_{k+1} - X_k) &= A^{-1}(b - A\cdot X_k) \end{aligned} \qquad (3.18)$$

Assuming R_0 to be an approximate inverse of matrix A, we get

$$X_{k+1} = X_k + R_0(b - A\cdot X_k)_{x_0=0 \text{ and } k=1,2,3.....}, \qquad (3.19)$$

where, A is the linear imaging operator, b is the set of observed low-resolution images and X_k is the kth approximation of the true scene. It can be easily confirmed by inspection of eq. (3.19) that when R_0 is the inverse of A the first iteration gives the exact solution. More interestingly, when A is not invertible but R_0 satisfies equality $AR_0A = A$, then again an exact solution is given by the first iteration in eq. (3.19) (assuming that a solution does exist). In that case R_0 is known to be a generalized inverse of A. In many inverse problems a suitable choice for R_0 is the operator adjoint to A multiplied by a small constant λ, that is $R_0 = \lambda A^*$. For such a choice of the operator R_0, eq. (3.19) is identical to an iterative minimization of the objective function $\Phi(X) = \|b - A X\|^2$ by the method of steepest descent.

The operator A^* also plays an important role in the proposed reconstruction scheme. If all the image degrading factors that contribute to the definition of the operator A, such as blur, geometrical warping *etc.* are neglected, keeping only translation and down-sampling (decimation) of the original HR image, then the operator A^* partially reverses the imaging process by putting LR image pixels back into the right places in the HR image from where they originated, leaving all other HR pixels equal to zero (depicted in fig. 3.3).

The operator R_0 in eq. (3.19) is defined as the superposition of the three consecutive operations, that is, translation and up-sampling of the LR frames, followed by interpolation of the HR image. More general warping of LR images is used when the simple translation motion model is not adequate. The interpolated high-resolution image is fed into the iteration scheme as the *first approximation image*, X_1. This image is then exposed to the same imaging conditions as the original (unknown) scene and is represented by the imaging operator A in eq. (3.15). As a result, a set of low-resolution images $b^{(1)}$ is generated. Since the image X_1 is only an approximation of the real scene, these new low-resolution images are different from the original low-resolution images b. The difference between the two sets forms the error vector that should eventually vanish when the reconstructed image converges to the real scene. The error vector is now reconstructed into high-resolution error image, using one of the interpolation techniques discussed in the previous sections, and added to X_1. The resultant image forms the *second approximation* X_2 of the real scene. The image, X_2 is now an input for the next iteration cycle. A new set of low-resolution images is generated and subtracted from the original set to form the new error vector, which is reconstructed and added to X_2. The resultant image is the *third approximation*, X_3. The process is repeated until convergence is achieved. This proposed technique is known as the ***Iterative-Interpolation Super-Resolution*** (IISR) [44, 45] reconstruction scheme. In this technique, each iteration only requires simple operations of frame shifting, up-sampling and convolution. As a result, the method is computationally efficient and highly suitable for parallel programming. The IISR technique can also be implemented on programmable hardware for real-world applications.

Figure 3.7 shows an example of IISR-reconstructed high-resolution image from fig. 3.4, 3.5 and 3.6. It is evident from fig. 3.7 that our technique successfully removed the artifacts that were visible in the interpolated images shown in fig. 3.6. It is worth mentioning that although the main purpose of the iteration procedure is to minimize artifacts appearing in the interpolated images, the process also contains implicit regularization features. The fundamental part of the reconstruction operator is the image interpolation technique. Both, the size and type of the

Fig. 3.7 IISR-generated high-resolution image

interpolation kernel as well as the number of iterations strongly affect the smoothness of the reconstructed image and control the stability of the process.

The quality of the reconstruction procedure scaled with the number of low-resolution images used in the reconstruction process. Greater the number of low-resolution images used, denser is the high-resolution composite grid, and better is the accuracy of interpolation and reconstruction. Note that for larger magnification factors, increasing the number of low-resolution images is necessary to maintain the same level of performance.

3.6 Similarity Measure

In order to quantify the fidelity of reconstruction, each reconstructed high-resolution image needs to be compared with the original scene; this is *similarity measure*. It also helps to monitor and evaluate the performance of the iterative reconstruction process. There are many similarity measures available in literature but none have been accepted universally as a standard when comparing two images. Some of the most popular are Normalized Cross-Correlation Ratio (NCCR), Direct Difference Error (DDE), Peak Signal-to-Noise Ratio (PSNR) and Root Mean Square Error (RMSE).

If an image is multiplied or added with a constant, the intensity level of the image changes but the spatial resolution of the image remains unchanged. In this research, the emphasis is on the spatial improvement of the reconstructed high-resolution image and therefore a modified version of RMSE has been utilized as a similarity measure. The modification was to ensure that the local spatial contents of images were compared and not their global brightness. Suppose $\boldsymbol{X}$ is the original high-resolution image and $\boldsymbol{X}'$ is the estimated high-resolution image, then the root mean square error, $\boldsymbol{E}_{rmse}$ is given as:

$$E_{rmse} = \frac{\sqrt{\frac{1}{IJ}\sum_{ij}(\tilde{X}_{ij} - \tilde{X}'_{ij})^2}}{\sigma_X}, \tag{3.20}$$

where,

$$\begin{aligned} \tilde{X} &= X_{ij} - \bar{X} \\ \tilde{X}' &= X'_{ij} - \bar{X}' \end{aligned}, \tag{3.21}$$

and

$$\sigma_X = \sqrt{\frac{1}{IJ}\sum_{ij}(X_{ij} - \bar{X})^2}. \tag{3.22}$$

PSNR is also calculated for each reconstructed high-resolution image and is given as:

$$E_{psnr} = 20 * \log_{10}\left(\frac{255}{E_{rmse}}\right). \tag{3.23}$$

3.7 Simulation Results

For the purpose of evaluating the performance of the proposed reconstruction scheme, several different test images (real and synthetic) were used. An example of successful super-resolution reconstruction using the proposed reconstruction scheme is presented in this section. The actual high-resolution test image (real imagery) is blurred with a Gaussian kernel of standard deviation 2 pixels. The LR images were artificially generated by randomly sub-sampling the blurred test image with the desired decimation ratio. The interpolation kernel used in the process of generating the high-resolution image is cubic-spline polynomial of order 6.

3.7.1 Noiseless LR Frames

In this particular example 10 LR images were used for the reconstruction and the sub-sampling ratio was 16. Figure 3.8 shows one of the ten LR images of the 'man' sequence. Please note that these LR frames were assumed to be noiseless.

Fig. 3.8 One of the 10 noiseless LR frames of the 'man' sequence [64 x 64]

The sequence of LR images is registered precisely relative to the reference LR frame. Once this is achieved, a high-resolution image grid is populated with pixels from low-resolution images by placing them at the appropriate grid-points according to the registration information. Figure 3.9 shows the high-resolution image grid for the 'man' sequence.

The *first approximation* of the high-resolution image, shown in fig. 3.10, is then estimated by interpolating the sparse grid shown in fig. 3.9. This *first approximation* image is now fed into the iterative reconstruction scheme to generate the final high-resolution image shown in fig. 3.11. The IISR technique converges and generates the final high-resolution image after 20 iterations only. By inspecting the final high-resolution image, fig. 3.11, one can clearly note that the deficiency of the reconstruction process seen in the form of periodic artifacts in the first approximation interpolated image is successfully removed.

Fig. 3.9 High-resolution composite grid image of the 'man' sequence [1024 x 1024]

Fig. 3.10 First approximation interpolated image of the 'man' sequence [1024 x 1024]

Fig. 3.11 IISR-generated high-resolution image from only 10 noiseless LR frames of the 'man' sequence [1024 x 1024] with an E_{rmse} = 0.14993 and E_{psnr} = 64.6131 dB

3.7.2 Noise Corrupted LR Frames

To check for the robustness of our reconstruction scheme, all the parameters used in the previous simulation are kept same, except in this case we introduce a noise level of 20dB into the LR frames. The various stages of reconstructing the high-resolution frame remain the same. Figure 3.12 shows one of the ten noise corrupted LR images of the 'man' sequence.

The first approximation of the high-resolution image is shown in fig. 3.13 and after 20 iterations; the final high-resolution image is estimated and is shown in fig. 3.14.

Fig. 3.12 One of the 10 noise corrupted LR frames of the 'man' sequence [64 x 64]

Fig. 3.13 Noise corrupted first approximation interpolated image of the 'man' sequence [1024 x 1024]

Fig. 3.14 IISR-generated high-resolution image from only 10 noise corrupted LR frames of the 'man' sequence [1024 x 1024] with an E_{rmse} = 0.17301 and E_{psnr} = 63.3696 dB

3.8 Quantitative Analysis

In order to compare and analyse the performance of all the interpolation techniques in reconstructing high-resolution images using the IISR scheme, a number of different high-resolution test images have been used in the simulation. For all images the RMSE tendencies were similar but the numbers differed in each one of them. In this section, quantitative analysis is presented for the super-resolution reconstruction of the 'man' sequence shown in previous section.

In Tables 3.1 and 3.2, we compare the image quality of the various interpolation techniques as a function of number of LR frames. In table 3.1, the LR frames (fig. 3.8) are assumed to be noiseless whereas in table 3.2 they are corrupted with a noise level of 20 dB (fig. 3.12). It clearly indicates that the improvement in the reconstruction quality or decrease of the RMSE value for each of the interpolation techniques is inversely proportional to the number of LR frames used.

Table 3.1 Comparison of reconstruction quality of interpolation techniques based on varying noiseless LR frames of the 'Man' sequence

Interpolation Techniques	Number of LR Frames			
	5	10	15	20
Nearest Neighbor	3.11457	10.2478	14.3594	18.1408
Linear	0.458858	0.210764	0.177949	0.155656
Cubic Polynomial: order 4	1.59841	0.153519	0.148448	0.140961
Cubic Polynomial: order 6	6.69829	0.149928	0.146297	0.139614
Gaussian: order 2	3.03074	0.169404	0.153711	0.147115
Gaussian: order 6	6.26985	0.149164	0.145088	0.138104
Gaussian: order 10	12.1345	0.147738	0.144277	0.137635
Truncated Sinc	47.7366	0.37388	0.275926	0.241462
Sinc Lanczos: degree 2	2.32884	0.153882	0.148138	0.140416
Sinc Lanczos: degree 3	9.72433	0.150709	0.147266	0.140588

Table 3.2 Comparison of reconstruction quality of interpolation techniques based on varying the number of noise corrupted LR frames of the 'Man' sequence

Interpolation Techniques	Number of LR Frames			
	5	10	15	20
Nearest Neighbor	2.8643	10.4806	15.0078	18.9061
Linear	0.465298	0.22478	0.191729	0.169404
Cubic Polynomial: order 4	1.3532	0.179433	0.167601	0.157011
Cubic Polynomial: order 6	6.16309	0.171794	0.162103	0.152532
Gaussian: order 2	3.08324	0.196791	0.177697	0.166279
Gaussian: order 6	6.05093	0.173366	0.161341	0.151416
Gaussian: order 10	11.733	0.170006	0.160193	0.149752
Truncated Sinc	44.3527	0.45666	0.319326	0.271369
Sinc Lanczos: degree 2	2.05446	0.178843	0.166752	0.155322
Sinc Lanczos: degree 3	10.3659	0.173986	0.162772	0.153341

Theoretically, if sufficient number of LR frames is available, each frame will give distinct information of the true scene or region of interest. A high-resolution image can then be reconstructed by mapping all the low-resolution pixels into the high-resolution frame. However, in practice, very often this ideal situation does not exist and it might not be possible to get a large number of low-resolution frames. The reason being, not all frames would have the subject or region of interest and also the subject may not be rigid in all the frames. With a large number of frames, imaging errors or atmospheric errors could also amplify causing erroneous reconstruction of the high-resolution image. In both, table 3.1 and 3.2, nearest-neighbor fails to perform even as the number of LR frames is increased. On the other hand, taking fewer frames will result in insufficient information to generate an approximate high-resolution image. This tendency can be seen in tables, 3.1 and 3.2, for all the interpolation kernels except nearest-neighbor at LR frames equal to 5. Therefore depending upon the nature of application an optimal number of LR frames should be used in order to generate a high-resolution image.

It is also evident from tables 3.1 and 3.2; a higher degree interpolation polynomial proves to be better than a polynomial of a lower degree. As expected, the nearest-neighbor, truncated Sinc, and linear interpolation kernels do not perform well as compared to the cubic spline, gaussian and lanczos window for Sinc. The data represented in tables 3.1 and 3.2, are generated from the 'man' sequence using sampling ratio of 16 with 20 iterations cycle.

Tables 3.3 - 3.4 show RMSE with varying number of iterations for both noiseless (fig. 3.8) and noise corrupted (20dB; fig. 3.12) LR frames respectively. The data represented in these tables, are generated from the 'man' sequence with a sampling ratio of 16 using only 10 LR frames. The table distinctly shows that gaussian interpolation kernel of order 10 generates the lowest RMSE results. The performance of the truncated Sinc, linear and nearest-neighbor is poor as compared to the other interpolation kernels. As mentioned earlier, tables 3.3 - 3.4 also prove that higher degree interpolation polynomials are better than their lower degree counterparts. The performance trend of all the interpolation kernels is almost similar in all the tables.

Tables 3.5-3.6 represent the data generated by taking RMSE as a function of varying sampling ratio for the 'man' sequence using 20 iterative cycles and only 10 LR frames. In table 3.5, the LR frames (fig. 3.8) are assumed to be noiseless whereas in table 3.6, they are corrupted with a noise level of 20 dB (fig. 3.12). It can be seen from the data that in some cases the value of RMSE increases as the sampling ratio is increased. This is due to the fact that as sampling ratio is increased, more information is lost during the initial process of simulating LR

Table 3.3 Comparison of reconstruction quality of interpolation techniques based on varying iterations and noiseless LR frames of the 'Man' sequence

Interpolation Techniques	Number of Iterations				
	0	5	10	15	20
Nearest Neighbor	0.671756	1.42485	2.13838	4.35488	10.2478
Linear	0.298197	0.220329	0.213569	0.211571	0.210764
Cubic Polynomial: order 4	0.316769	0.171747	0.157746	0.154497	0.153519
Cubic Polynomial: order 6	0.319034	0.172327	0.156652	0.151957	0.149928
Gaussian: order 2	0.37388	0.166077	0.161354	0.164624	0.169404
Gaussian: order 6	0.315094	0.170486	0.155179	0.150845	0.149164
Gaussian: order 10	0.317821	0.170826	0.154991	0.149995	0.147738
Truncated Sinc	0.339865	0.21187	0.211531	0.251259	0.37388
Sinc Lanczos: degree 2	0.323435	0.169561	0.156368	0.154133	0.153882
Sinc Lanczos: degree 3	0.322944	0.173038	0.157201	0.152601	0.150709

Table 3.4 Comparison of reconstruction quality of interpolation techniques based on varying iterations and noise corrupted LR frames of the 'Man' sequence

Interpolation Techniques	Number of Iterations				
	0	5	10	15	20
Nearest Neighbor	0.674621	1.43206	2.16567	4.41021	10.4354
Linear	0.301279	0.230817	0.226703	0.225501	0.224968
Cubic Polynomial: order 4	0.320731	0.188545	0.179451	0.178067	0.178999
Cubic Polynomial: order 6	0.323626	0.187476	0.175793	0.173449	0.172176
Gaussian: order 2	0.375842	0.18474	0.185059	0.191375	0.197618
Gaussian: order 6	0.319857	0.186134	0.174974	0.173004	0.173035
Gaussian: order 10	0.323094	0.18604	0.174183	0.171365	0.16856
Truncated Sinc	0.346529	0.225485	0.232701	0.286627	0.455986
Sinc Lanczos: degree 2	0.327417	0.186517	0.17821	0.178809	0.180125
Sinc Lanczos: degree 3	0.327645	0.188467	0.17638	0.173245	0.173268

Table 3.5 Comparison of reconstruction quality of interpolation techniques based on varying sampling ratio and noiseless LR frames of the 'Man' sequence

Interpolation Techniques	Sampling Ratio			
	8	12	16	20
Nearest Neighbor	7.99961	9.0566	10.2478	10.9996
Linear	0.16474	0.167131	0.210764	0.179999
Cubic Polynomial: order 4	0.0874763	0.0369791	0.153519	0.106596
Cubic Polynomial: order 6	0.08256	0.0403187	0.149928	0.113179
Gaussian: order 2	0.170384	0.0781008	0.169404	0.134025
Gaussian: order 6	0.0807436	0.032459	0.149164	0.107406
Gaussian: order 10	0.07958	0.0340558	0.147738	0.111083
Truncated Sinc	0.166821	0.218574	0.37388	0.51656
Sinc Lanczos: degree 2	0.0836392	0.0332509	0.153882	0.110628
Sinc Lanczos: degree 3	0.0807327	0.0375689	0.150709	0.119075

images and then at the reconstruction phase not all of this information is recovered in the high-resolution estimate causing a rise in relative error.

Figures 3.15 - 3.16 show plots between RMSE and the number of iterations, used in the proposed IISR scheme, as a function of LR frames taken during the initial reconstruction process. The data represented in the plots are generated from the 'man' sequence using 10 LR frames, sampling ratio of 16 with 20 iterations cycle.

Although the tendency of the graph looks similar in both the plots, fig. 3.15 was generated considering noiseless LR frames whereas fig. 3.16 considered LR frames

Table 3.6 Comparison of reconstruction quality of interpolation techniques based on varying sampling ratio and noise corrupted LR frames of the 'Man' sequence

Interpolation Techniques	Sampling Ratio			
	8	12	16	20
Nearest Neighbor	8.15911	9.23234	10.395	11.0522
Linear	0.184486	0.184644	0.224677	0.196194
Cubic Polynomial: order 4	0.128865	0.0990529	0.179259	0.140744
Cubic Polynomial: order 6	0.120211	0.0931465	0.172829	0.141838
Gaussian: order 2	0.203512	0.129347	0.197686	0.169639
Gaussian: order 6	0.120255	0.0920756	0.172967	0.13894
Gaussian: order 10	0.117811	0.0899574	0.170626	0.139185
Truncated Sinc	0.32715	0.345145	0.463106	0.577806
Sinc Lanczos: degree 2	0.128037	0.0983905	0.179729	0.144
Sinc Lanczos: degree 3	0.11918	0.0924772	0.172918	0.145989

corrupted by a noise level of 20 dB. These plots provide a significant overview as to how each of the interpolation techniques fair in the race for achieving super-resolution. As mentioned earlier and can also be seen in fig. 3.15 - 3.16 that a higher degree interpolation kernel proves to be better than a kernel of lower degree. Gaussian kernel of order 10 is found to be the best interpolation kernel since it achieves the lowest relative error and generates a high-resolution image very close to the original test image. In both plots (fig. 3.15 - 3.16), the graph drops steeply between 1-to-4 iterations indicating that most of the major improvement, i.e. removal of artifacts, occurs during the initial first few iterative cycles. The improvement is more gradual in the later cycles, indicated by the flattening of the graph. Gaussian interpolation kernel of order 2 shows a peculiar characteristic in the plots, instead of flattening out gradually the graph starts to rise again. This indicates that the image quality is deteriorating after about 8-to-10 iterations only. Therefore, for gaussian kernel of order 2, the optimum number of iteration cycles is 10 in this particular example of super-resolution of the 'man' sequence.

Super-resolution image reconstruction plays a significant role in various fields of both, civil and military domains, such as text recognition, target recognition, medical imaging and many more. Although it is very important for super-resolution to be highly accurate but depending upon its field of application, the computational time taken to generate such accurate results has to be low and acceptable in that field. For example, in real time video surveillance or target detection systems, computational time is of great importance and hence requires super-resolution technique with high accuracy and low computational cost as compared to astronomical imaging or text recognition where computational time is not a constraint. Therefore in order to evaluate our proposed IISR technique in terms of computational time taken by each of the interpolation kernels, three different high-resolution test images – Lena (size - 512 x 512), Man (size - 1024 x 1024) and ISO (size - 2048 x 2048) were used for generating the data. A sampling

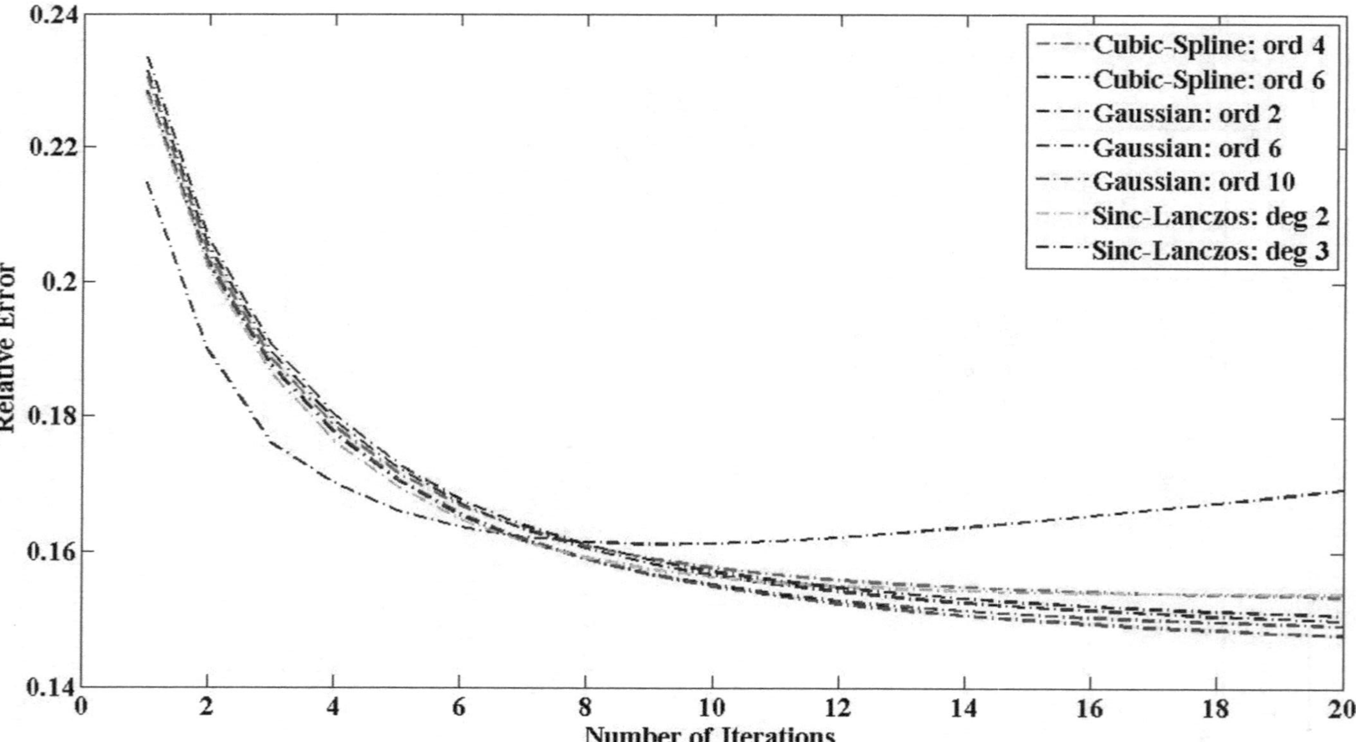

Fig. 3.15 Relative error or RMSE of the reconstructed high-resolution image of the 'man' sequence by all the interpolation kernels (noiseless)

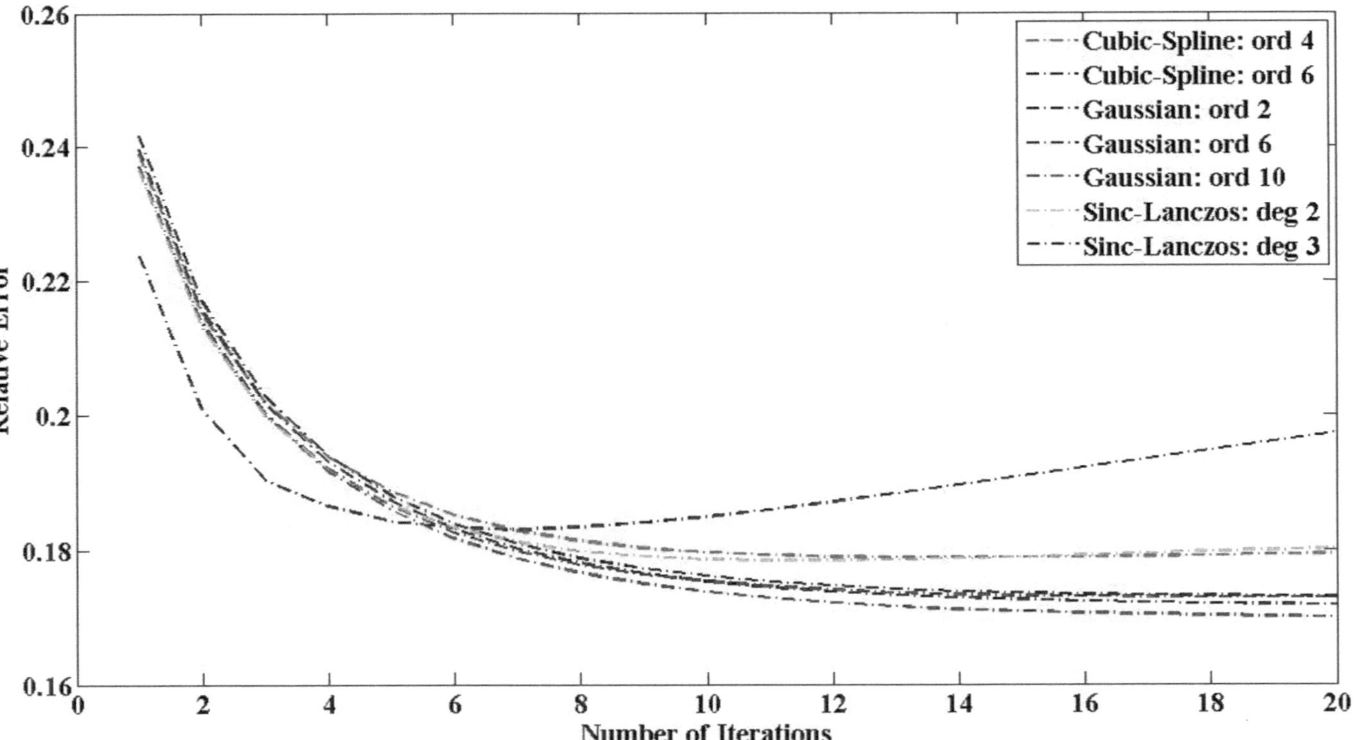

Fig. 3.16 Relative error or RMSE of the reconstructed high-resolution image of the 'man' sequence by all the interpolation kernels (noise corrupted)

ratio of 16 was used to generate 10 randomly translational shifted LR frames from each high-resolution test image. IISR image reconstruction was performed using all interpolation kernels on each of the three sets of LR frames and the estimated high-resolution images were then generated after 20 iterations. Tables 3.7 - 3.8 represents the computational time analysis data of the IISR system based on noiseless and noise corrupted (20 dB) LR frames respectively. The graphical representation of these tables is given by figs. 3.17 - 3.18 respectively. The time analysis data was generated on an Intel Pentium M processor (1.40 GHz) notebook with 1.00 GB of RAM.

It is important to note that the data tabulated in tables 3.7 - 3.8 is the total computational time taken by the IISR system to generate the final high-resolution

Table 3.7 Computational time analysis of IISR system based on noiseless LR frames

Interpolation Techniques	Time (in secs)		
	Lena (512 x 512)	Man (1024 x 1024)	ISO (2048 x 2048)
Nearest-Neighbor	2.1398	7.6378	69.2339
Linear	2.0804	8.111	90.5327
Cubic Polynomial: order 4	2.6241	10.0989	83.0398
Cubic Polynomial: order 6	3.5235	13.8185	152.7094
Gaussian: order 2	2.08	8.1219	70.41
Gaussian: order 6	3.5801	13.8207	118.2392
Gaussian: order 10	6.44	26.1593	159.9797
Truncated Sinc	2.5872	10.0802	106.1028
Sinc Lanczos: degree 2	2.5814	10.088	90.3638
Sinc Lanczos: degree 3	3.5068	13.8181	105.9359

Table 3.8 Computational time analysis of IISR system based on noise corrupted LR frames

Interpolation Techniques	Time (in secs)		
	Lena (512 x 512)	Man (1024 x 1024)	ISO (2048 x 2048)
Nearest-Neighbor	2.7031	12.4124	90.5229
Linear	2.8961	14.1946	88.4304
Cubic Polynomial: order 4	3.5388	19.8057	100.7633
Cubic Polynomial: order 6	4.7323	30.632	120.0999
Gaussian: order 2	2.8285	18.2192	142.9922
Gaussian: order 6	4.7322	30.4941	116.8328
Gaussian: order 10	9.006	56.7097	182.0827
Truncated Sinc	3.4628	14.72	96.7034
Sinc Lanczos: degree 2	3.4887	15.9169	103.4364
Sinc Lanczos: degree 3	4.7591	18.6307	112.3867

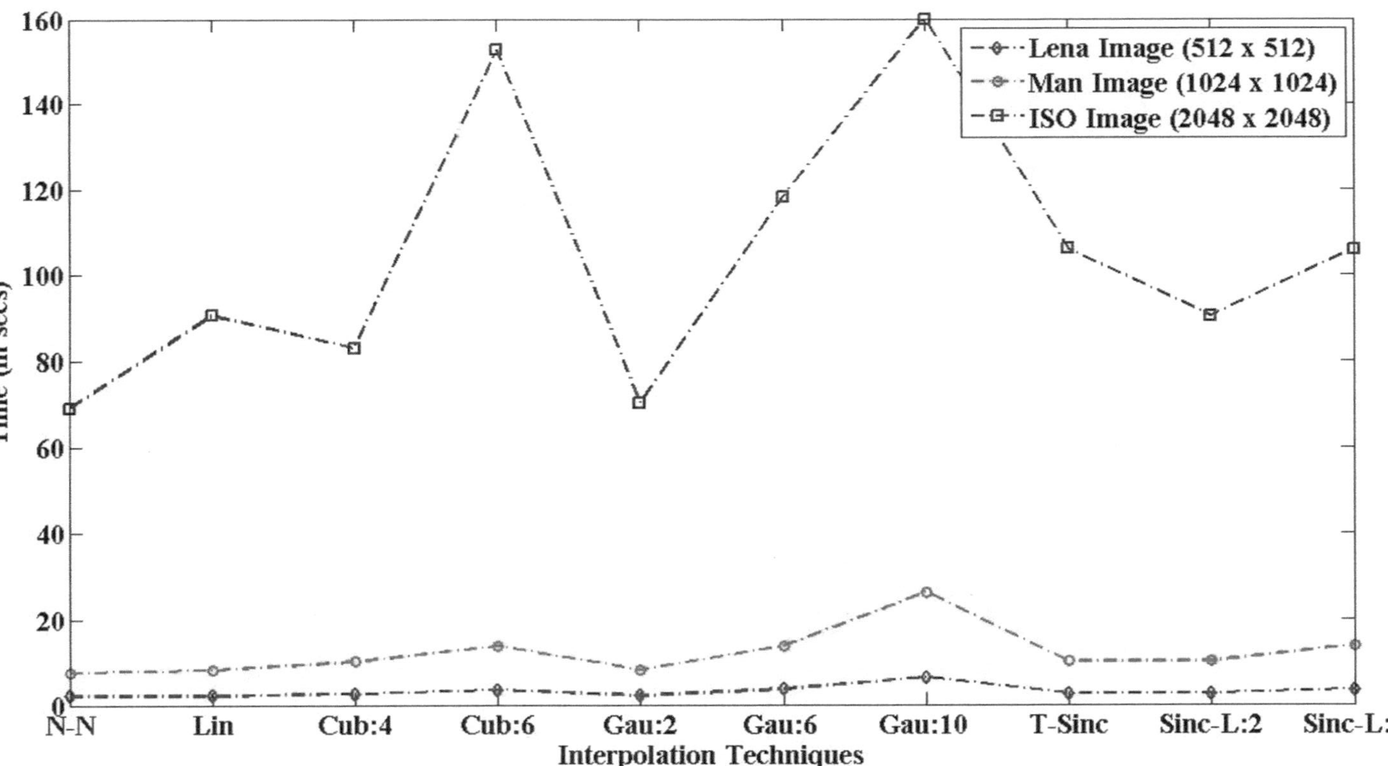

Fig. 3.17 Computational time analysis of IISR system (noiseless super-resolution)

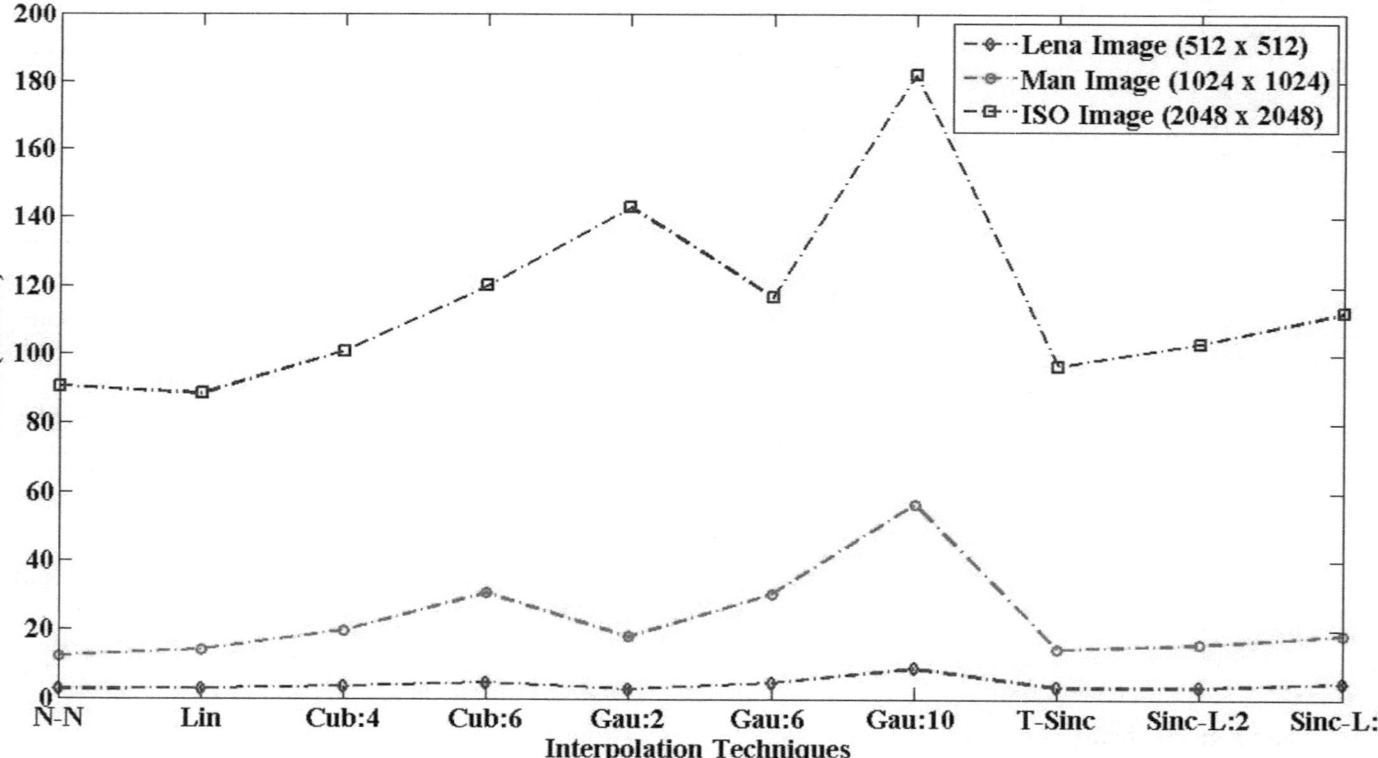

Fig. 3.18 Computational time analysis of IISR system (noise corrupted super-resolution)

image using the specified number of iterations and assuming that the LR frames and their registration information was already available.

From fig. 3.15 - 3.16 it is very much evident that the higher degree interpolation kernels are more accurate than the lower degree ones. On the other hand, the computational time analysis data shown in fig. 3.17 - 3.18 clearly changes this fact for applications of super-resolution where time constraints are significant. For generating high-resolution images of size 512 x 512, all the interpolation kernels are acceptable in terms of computational speed but after considering the accuracy of results, cubic-spline, gaussian and Sinc-lanczos kernels are the suitable ones. As the resultant final image size is increased to 1024 x 1024 or even 2048 x 2048, the list of acceptable interpolation kernel decreases depending upon the accuracy of results and computational speed. From the experimental data presented in this section, gaussian kernel of order 6 and cubic spline polynomial of order 4 certainly qualify as the best tradeoff between accuracy and computational speed.

3.9 Summary

In this chapter, an innovative and hybrid reconstruction scheme has been proposed for achieving super-resolution from a sequence of geometrically warped, aliased, and under-sampled low-resolution images. The proposed reconstruction scheme uses interpolation techniques to produce the *first approximation* of high-resolution image and then employs an iterative approach to generate the final solution.

Performance of the IISR technique is significantly dependent on the type of interpolation kernel used in reconstruction process. Several different interpolation techniques have been implemented and tested. An extensive quantitative analysis was conducted and several different test high-resolution images were used to evaluate the fidelity of the IISR system. All computer simulations and data analysis were done using both noiseless and noise corrupted LR frames. The tendency of the reconstruction scheme remained same in both the cases thereby confirming the robustness of IISR system. As expected, the higher degree interpolation methods were more accurate, leading to better reconstruction. The price for better accuracy, however, was extended computational time. Considering this fact in terms of practicality, Gaussian kernel of order 6 and cubic spline polynomial of order 4 or 6 were the most promising in the tradeoff between the accuracy and the computational speed.

The quality of the reconstruction procedure scaled with the number of LR images used in the process. As the number of LR frames is increased, more densely populated the composite high-resolution grid gets and greater is the accuracy of the reconstruction scheme. Note that for larger magnification factors, increasing the number of LR frames is necessary to maintain the same level of performance. In most cases, about 10 LR frames were sufficient for magnification factors as large as 20.

The proposed IISR system successfully handles the under-determined problem of super-resolution and requires a relatively small number of low-resolution images for accurate reconstruction. Good results were obtained with only 10 LR frames for magnification factors as large as 20. This is important for practical applications, because if a large number of LR images were required the accumulation of errors would impede the reconstruction accuracy. Each iteration in the IISR scheme only requires simple operations of frame shifting, up-sampling and convolution. As a result, the method is computationally efficient and is promising to be applicable for real-time processing by using programmable hardware.

Chapter 4
Optimization Approach to Super-Resolution Image Reconstruction

4.1 Introduction

The process of super-resolution is an inverse problem of estimating a high-resolution image from a sequence of observed, low-resolution images and it is now widely known to be intrinsically unstable or "ill-conditioned". The common feature of such ill-conditioned problems is that the small variations in the observed images can cause (arbitrary) large changes in the reconstruction. This sensitivity of the reconstruction process on the input data errors may lead to the restoration errors that are practically unbounded. The important part of super-resolution process is thus to modify the original problem in such a way that the solution is meaningful and a close approximation of the true scene but, at the same time, it is less sensitive to errors in the observed images. The procedure of achieving this goal and to stabilize the reconstruction process is known as *Regularization*.

Many super-resolution techniques [8, 10, 16] are based on the optimization approach. We re-examine the approach to super-resolution using optimization techniques by reformulating the problem in terms of a regularized optimization procedure and implement an iterative conjugate gradient method for finding the minimum of the resulting objective function. The role of regularization term on the accuracy of the super-resolution reconstruction is also investigated. For evaluating the performance of the IISR technique, the full-solution of IISR is compared with the one generated by the optimization technique [46].

4.2 Well-Posed vs Ill-Posed

There are many ways of explaining well-posed and ill-posed problems. For example,

$$AX = b\,, \tag{4.1}$$

where $\boldsymbol{A}$ is known. Now, if $\boldsymbol{b}$ is determined by $\boldsymbol{X}$, this is a well-posed problem whereas if X has to be determined from b, it's an inverse or ill-posed problem. The latter relates to super-resolution as explained in the introduction.

V. Bannore: Iterative-Interpolation Super-Resolution Img. Recons., SCI 195, pp. 51 – 76.
springerlink.com

A problem whose solution exists, is unique and depends on the data continuously is known as a *well-posed* problem as defined by Hadamard [47] in 1902. On the contrary, *ill-posed* problem is the one which disobeys the rules defined by Hadamard. In addition, as the solution of the ill-posed problem depends in a discontinuous fashion on the data, small errors such as round-off and measurement errors, may lead to a highly erroneous solution. The solution for an ill-posed problem is unstable and extremely sensitive to fluctuations in the data and other parameters. The classical example of an inverse and ill-posed problem is the Fredholm integral equation of the first kind, where, k is the kernel and g is the right-hand side.

$$\int_a^b k(t,s)\, f(s) dx = g(t) \quad . \tag{4.2}$$

Both of these parameters, $\boldsymbol{k}$ and $\boldsymbol{g}$ are known, while $\boldsymbol{f}$ is the unknown function to be computed [48]. The theory on ill-posed problems is quite extensive and well developed. Engl [49] conducted a survey on a number of practical inverse problems in various applications such as computerised tomography, heat conduction, and inverse scattering problems. Inverse problems are also seen in various other fields, such as, medical imaging, astronomy, and many more. Ill-conditioning of inverse problems has always attracted a great deal of interest in research.

For many decades, it has been known that the best way to analyse a scientific problem is through its mathematical analysis. The most common analytical tool used in the case of ill-posed problems is *Singular Value Decomposition (SVD).* This tool helps in diagnosing whether or not the singular values of a matrix are zero or decaying slowly towards zero (a number is so numerically small that due to the round off error it is rounded to zero). The SVD for a matrix $\boldsymbol{A}$ of dimension m by n where $m \geq n$, is given by:

$$A = USV^T \Rightarrow A = \sum_{i=1}^{z} u_i s_i v_i^T \quad . \tag{4.3}$$

For the above decomposition, $\boldsymbol{U}$ $(u_1.....u_z)$ is an m by m and $\boldsymbol{V}^{\mathbf{T}}$ is the transpose of matrix $\boldsymbol{V}$ $(v_1.....v_z)$ which is n by n. The matrix S is a diagonal matrix containing the non-negative singular values of $\boldsymbol{A}$ arranged in descending order. The matrix $\boldsymbol{U}$ and $\boldsymbol{V}$ are orthogonal and their columns are orthonormal. The columns u_i and v_i of $\boldsymbol{U}$ and $\boldsymbol{V}$ are known as the left and right singular vectors. Also, for certain applications, as the dimension of matrix $\boldsymbol{A}$ increases, the numerical value of the singular values in $\boldsymbol{S}$ gradually decreases towards zero which causes more oscillations in the left and right singular vectors. Greater the number of singular values in $\boldsymbol{S}$ tending to zero, the more singular is matrix $\boldsymbol{A}$ making it more ill-conditioned. Thus, SVD gives a good approximation on the ill-conditioning of the system.

Another easier way of testing a system for ill-conditioning is by computing the *condition number* of the matrix. The condition number can be defined as a ratio of the maximum and minimum singular values of the matrix in consideration, for example, $\boldsymbol{A}$ from eq. (4.1). A high condition number points to an ill-posed problem, whereas a low condition number points to a well-posed problem. If $\boldsymbol{A}$ is an m by n matrix:

$$condition(A) = \left| \frac{\sigma_{max}(A)}{\sigma_{min}(A)} \right|_{euclidean-norm}, \tag{4.4}$$

where, σ_{max} and σ_{min} represent the maximum and minimum singular values of matrix $\boldsymbol{A}$. With ill-posed problems, the challenge is not of computing a solution, but computing a unique and stabilized solution. Thus, an ill-conditioned system requires an intelligent method of mathematical computation to generate a meaningful solution, rather than the usual computational methods.

Referring to spatial model of super-resolution, eq. (3.15) renamed to eq. (4.5), may have many solutions or, due to noise, may not have any solution whatsoever.

$$\underline{b} = A\underline{X} + \underline{e} . \tag{4.5}$$

An approximate least square solution may be then obtained by minimization of the error between the actually observed and the predicted LR images. Thus, the objective function $\boldsymbol{\Phi}$ is given by the following expression:

$$\Phi(X) = \min \left\| b - AX \right\|_2^2 , \tag{4.6}$$

where, $\| \ \|_2$ is the Euclidean or L_2 norm. As mentioned before, solutions of inverse problems are very often unstable due to their sensitivity to small perturbations of the data. Mathematically, this sensitivity occurs due to the existence of very small singular values (as defined by the singular value decomposition) in the spectrum of the matrix $\boldsymbol{A}$. The matrix $\boldsymbol{A}$ is therefore, singular in nature and highly ill-conditioned. There is no uniqueness and stability in the solution for eq. (4.6). Therefore to make the above equation well-conditioned (as per Hadamard criteria), another term is added to eq. (4.6) known as the *Regularization Term.* Most of the inverse problems (such as super-resolution) are ill-posed and the solution is highly sensitive to the data. The solution can vary tremendously in an arbitrary manner with very small changes in the data. The solution to eq. (4.6) would be highly sensitive and noise contaminated. The regularization term takes control of the ill-conditioned nature of the problem. The aim of this term is to make the solution more stable and less noise contaminated. The term also attempts to converge the approximate solution as close as possible to the true solution. The modified version of eq. (4.6) is:

$$\Phi(X) = \min\left[\left\|\underline{b} - A\underline{X}\right\|_2^2 + \lambda\left\|Q\underline{X}\right\|_2^2\right]. \tag{4.7}$$

In eq. (4.7), the parameter $\lambda > 0$, is known as the regularization parameter and $\boldsymbol{Q}$ is a regularization / stabilization matrix. In [Hanke-Hansen 1993], the stabilization matrix is referred to as the regularization operator. The regularization operator if given by an identity matrix ($\boldsymbol{Q}$ = I), the regularization term is of *standard form* whereas when $\boldsymbol{Q} \neq$ I, the term is in the *general form*. When treating problems numerically, it is easier to use the standard form rather than the general form as only one matrix, $\boldsymbol{Q}$, needs to be handled. In practical applications, however, it is recommended that the general form of the regularization term should be used.

There is no unique procedure for constructing the regularization term. In the Bayesian type of approach, eq. (4.7) is an example of the MAP estimator where the regularization term incorporates *a priori* knowledge about the problem. The method, however, offers little practical guidance of how to construct the term. In the context of super-resolution, a popular choice for the matrix $\boldsymbol{Q}$ is a discrete approximation of the Laplacian operator that penalizes large variations in the estimated image and enforces its smoothness. For the purpose of comparison and experimentation, we have implemented four different forms of regularization matrix that are based on various discrete representations of Laplacian and Biharmonic operators. They are shown in fig. 4.1 as convolution kernels.

The regularization term aims at filtering out the noise that contaminates the image and also makes it smoother. The regularization term can also include *a priori* information of the true solution which facilitates the minimization process to converge as close as possible. The regularization parameter controls the measure of smoothness in the final solution of eq. (4.7). It is critical to choose the regularization parameter best suited to the particular application in which it is involved. If the regularization parameter is too small, it brings back the risk for the solution being unstable and susceptible to noise amplification, causing the approximate solution to be far from converging with the true solution. On the other hand, if the parameter is too large, the regularization term will have a dominating effect on the solution making it too smooth and some information may be lost. Hence, there needs to be a proper balance of smoothness and preservation of information when regularization is implemented.

4.3 Estimating λ, the Regularization Parameter

Being the most critical part of regularization term, one has to carefully choose the appropriate technique based on their applications and expected results. Also, the choice of the optimal value of the regularization parameter may strongly influence the fidelity of the reconstruction process. The regularization parameter also depends on the properties of $\boldsymbol{b}$, $\boldsymbol{A}$, $\boldsymbol{X}$ and noise (eq. 4.7). The parameter should balance the regularization and perturbation error in the computed solution. Over the years, many techniques have been proposed and discussed in relation to

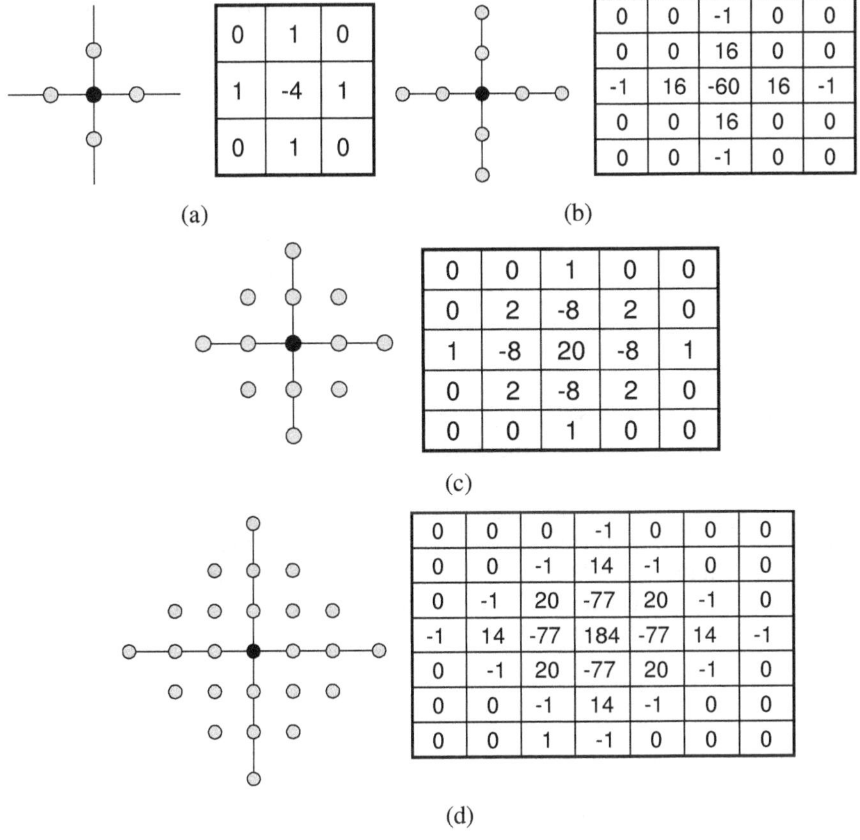

Fig. 4.1 Regularization/Stabilization Matrix (***Q***): (a) 2D Laplacian Operator (***Q*** = 4-neighbor). (b) 2D Laplacian Operator (***Q*** = 8-neighbor). (c) 2D Biharmonic Operator (***Q*** = 12-neighbor). (d) 2D Biharmonic Operator (***Q*** = 24-neighbor).

estimating the regularization parameter [50-53]. The techniques discussed fall into two categories – one which require knowledge of error and the ones which do not require knowledge of error.

4.3.1 Methods Which Require Error Knowledge

4.3.1.1 The Discrepancy Principle

In practical scenarios, considering eq. (4.1) and (4.6), the right–hand side term, $\boldsymbol{b}$, is never free from errors and contains various types of errors. Therefore, $\boldsymbol{b}$ can be written as $\boldsymbol{b} = \boldsymbol{b}_{true} + \boldsymbol{e}$, where $\boldsymbol{e}$ is the errors and $\boldsymbol{b}_{true}$ is the actual unperturbed right-hand side. Now, as per the discrepancy principle [54], the regularization parameter is chosen such that the residual norm of the regularized solution is equal to the norm of the errors.

$$\left\| b - AX_{reg} \right\|^2 = \left\| e \right\|^2 . \tag{4.8}$$

If there is a rough estimate of the error norm, the discrepancy principle can be used to estimate a good regularization parameter. Unfortunately, in the practical world the knowledge about the error norm is not available and can be erroneous. Such data can lead to wrong estimations of the regularization parameter, thereby generating an unstable final solution.

4.3.1.2 λ-Curve

In the context of super-resolution reconstruction of images, the ultimate measure of the fidelity of reconstruction is the difference between the original test image and its attempted high-resolution reconstruction. More precisely, the RMSE of the two images is a good choice for the quantitative description of the accuracy of reconstruction. When plotted as a function of λ, the resulting curve is expected to go through a minimum. The point of minimum RMSE value is then selected as the optimal value of the regularization parameter λ. This is illustrated in fig. 4.2.

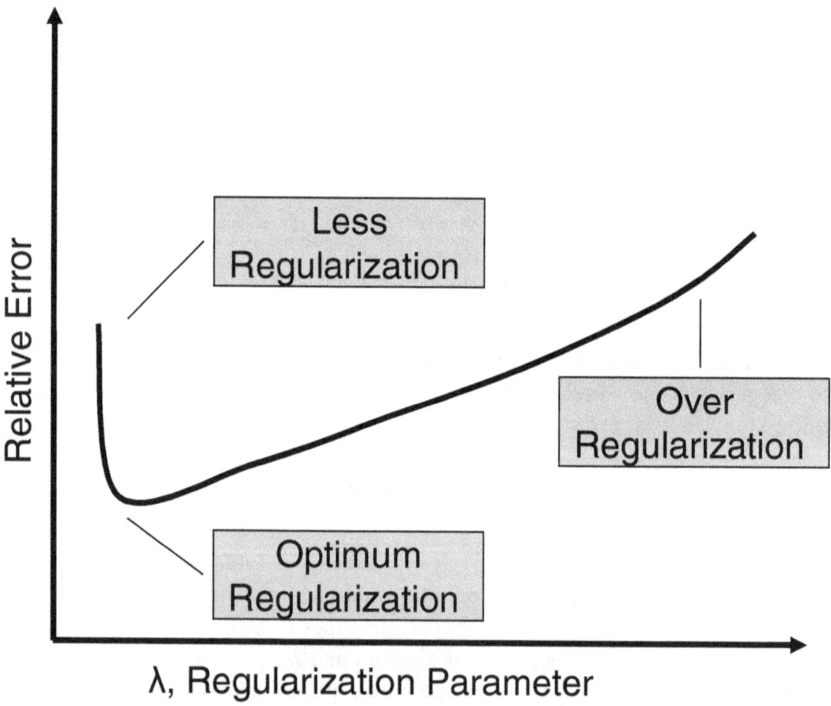

Fig. 4.2 A Generic λ-curve graph

The proposed λ-curve method is adequate for experimentations with simulated imagery, however, in practice the original scene is unknown and hence error knowledge is not available, therefore other methods have to be considered.

4.3.2 *Methods Which Do Not Require Error Knowledge*

4.3.2.1 Generalized Cross-Validation (GCV)

GCV is one of the most popular methods used for estimating the regularization parameter [55]. It is based on the statistical cross-validation technique. In GCV, if a random element, $\boldsymbol{b}_k$, is left out of $\boldsymbol{b}$, then the estimated regularized solution should be able to predict the missing element, $\boldsymbol{b}_k$. The regularization parameter is chosen as the one which minimizes the prediction error and is independent of the orthogonal transformation of $\boldsymbol{b}$, [56]. In this technique, no knowledge of the error norm is required. The GCV function is given as:

$$GCV = \frac{\left\| b - AX_{reg} \right\|^2}{\tau^2} \tag{4.9}$$

In eq. (4.9), the numerator is the squared residual norm and the denominator is the square of effective number of degrees of freedom. Although computation of regularization parameter using GCV technique works for many applications, it should also be noted that GCV may have a very flat minimum [57], making it difficult to locate numerically thereby failing to compute the correct λ.

4.3.2.2 L-Curve Criterion

The L-curve criterion proposed in [50, 58] was inspired from the graphical analysis discussed in [59]. The L-curve method directly exploits the competition between the two terms in eq. (4.7): the residual fitting error $\|\boldsymbol{b} - \boldsymbol{AX}\|^2$ and the stabilization term $\|\boldsymbol{QX}\|^2$. The values of these terms are calculated for $\boldsymbol{X}$ that minimizes the objective function $\boldsymbol{\Phi}$ for a given value of the parameter λ. This curve, when plotted on a log-log scale for a range of λ values, takes the shape which resembles the alphabet 'L' and hence the name, L-Curve (see fig.4.3 for illustration). This is the most powerful graphical tool for analysis as it shows the relationship between both the terms of eq. (4.7). The flat part of the L-curve corresponds to a solution dominated by the regularization errors whereas the vertical part indicates that the solution is dominated by perturbation errors. It was then argued [50, 58] that the optimal value of the parameter λ is at the corner of the curve where the curvature is at its maximum since it marks the optimal balance between minimizing the regularization error and the residual fitting error in the solution. This is the L-curve criterion.

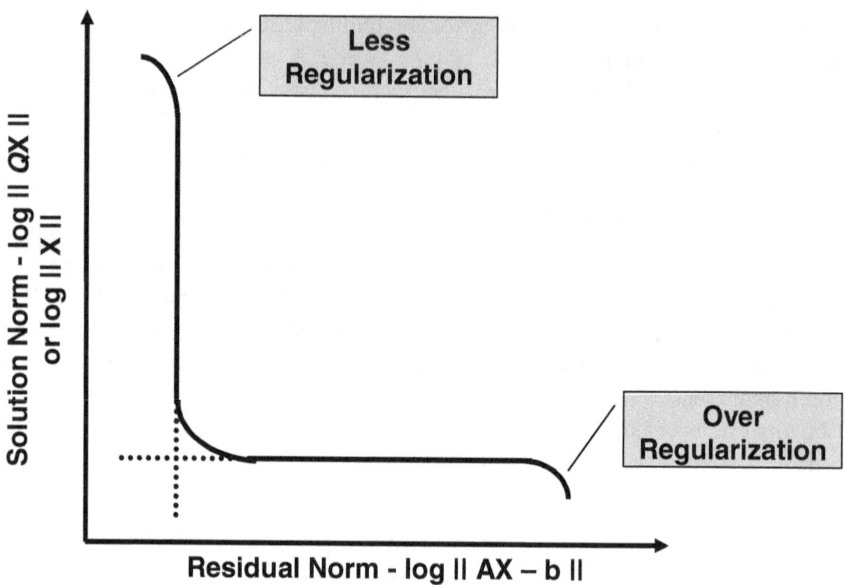

Fig. 4.3 A Generic L-curve graph

The corner of the L-curve is the point of maximum curvature, **κ** (kappa) [50]:

$$\kappa_\lambda = \frac{\hat{\rho}'\hat{\eta}'' - \hat{\rho}''\hat{\eta}'}{[(\hat{\rho}')^2 + (\hat{\eta}')^2]^{3/2}}, \tag{4.10}$$

where, ′ denotes differentiation and **ρ** and **η** are given as:

$$\begin{aligned} \rho_\lambda &= \log\left\| A\underline{X}_\lambda - b \right\|_2 \\ \eta_\lambda &= \log\left\| Q\underline{X}_\lambda \right\|_2 \end{aligned} \quad . \tag{4.11}$$

It is important to note that the trend of the L-curve and the point of optimum balance is not a generalization and can fluctuate depending upon (and is not limited to) the choice of regularization matrix, nature of data, field of application and even the type of norm used for minimization purposes. As mentioned earlier, although λ-curves are not practically feasible, however in a controlled environment, relative error is known or can be calculated, and one could employ λ-curves to validate the regularization parameter estimated by the L-curve criterion.

4.4 Regularization Techniques

Regularization is an intelligent technique for computing a solution for an ill-posed problem. The main aim of this term is to make sure that the final solution is smooth and regularized with respect to the input data. It also makes sure that the final solution is less contaminated with errors and noise components. In the process of achieving this, the regularization term filters out the high-frequency components, thereby giving a smooth final solution. In the field of image restoration or super-resolution, a smooth approximate solution might not solve the purpose of being an appropriate solution. The high-frequency components filtered out by the regularization technique relate to the edges and discontinuities in the image. These components hold significant value in the field of image restoration. As seen in section 4.2, the singular values of the matrix $\boldsymbol{A}$, are of critical importance, as they relate to the high-frequency components. Thus, if there are too many small singular values (which can decay to zero), then the information relating to these is lost and only the information related to the large values is recoverable. There are various techniques for computing a regularized solution for ill-posed problems.

4.4.1 Tikhonov Regularization

Tikhonov regularization [60-62] was first introduced in 1963 and is defined as:

$$X_{reg.} = \min\left[\|b - AX\|_2^2 + \lambda\|QX\|_2^2\right]. \tag{4.12}$$

$\lambda > 0$, is known as the regularization parameter and $\boldsymbol{Q}$ is a regularization/stabilization matrix. The regularization matrix can be $\boldsymbol{Q} = \mathrm{I}$ or $\boldsymbol{Q} \neq \mathrm{I}$ where I is an identity matrix. It is recommended to consider the regularization matrix as unequal to the identity matrix, [52]. It should be noted that since the regularization matrix can also contain *a priori* knowledge, greater care must be taken in its selection. The regularization parameter is also of great importance as it is a trade-off between the smoothness and the accuracy of the solution. Eq. (4.12) can be also written as:

$$(A^T A + \lambda Q^T Q) X_{approx} = A^T b. \tag{4.13}$$

From eq. (4.13), it is evident how the regularization term manages to regularize the solution. It is also evident how the proper or improper selection of λ and $\boldsymbol{Q}$ can lead to a good or bad approximation. A high value of λ diverts the solution to be very smooth, suppressing the high-frequency components even though the system has been regularized. Although Tikhonov regularization seems to be a straight forward technique, it has a high-computational cost and requires a lot of storage space when used in large-scale problems. Thus, this technique is more suitable to small-scale problems as compared to large-scale problems.

4.4.2 Maximum Entropy Method

The Maximum Entropy technique [63] of regularization is often used in astronomical image reconstruction. This technique is also known to preserve point edges in the estimated image, which makes it promising in the field of astronomical image restoration. The maximum entropy regularization term [64] is given as:

$$S(X) = \lambda^2 \sum_{i=1}^{z} x_i \log(w_i x_i) \quad , \tag{4.14}$$

where, $\boldsymbol{x_i}$ are the positive elements of vector $\boldsymbol{X}$ and $\boldsymbol{w_i}$ are weights $(w_1 \ldots\ldots w_z)$. The above given function is negative of the entropy function. Therefore, eq. (4.12) can now be written as:

$$X_{ME} = \left\| b - AX \right\|_2^2 + S(X) \quad . \tag{4.15}$$

The estimated solution from maximum entropy regularization is quite consistent as it is not related to the missing information of the right-hand side to a great extent. The major drawback of this technique is that it's computationally intensive to allow for it to be applicable in real-time processes.

4.4.3 Conjugate Gradient (Iterative Regularization)

Conjugate gradient is one of the most commonly used numerical algorithms for symmetric positive definite systems. It is also known as the oldest and best known non-stationary method. Conjugate gradient can be computed as a direct method much such as Tikhonov and maximum entropy but it proves to be much more efficient if it is used as an iterative method. Direct methods fail to perform when it comes to large-scale problems or huge sparse matrices, where only iterative technique comes to the rescue. The iterative conjugate gradient method can successfully compute solutions for large scale problems. Since the iterative method utilizes the property of matrix-vector multiplications between huge sparse matrices and vectors, computational time decreases and storage requirements for such matrices and vectors decreases tremendously. These advantages make iterative conjugate gradient regularization technique more favourable when compared with others. The iterative method generates successive approximations of the solution and their residuals. The conjugate gradient for a set of unregularized normal equations, $\boldsymbol{A\,X = b}$, is given as:

$$A^T A X = A^T b \quad . \tag{4.16}$$

It is seen that for eq. (4.16), the low-frequency components of the estimated solution converge faster than the high-frequency components [58]. The iterative conjugate gradient technique generates $\boldsymbol{X_K}$ estimated solutions and calculates the residuals for each $\boldsymbol{K}$, where $\boldsymbol{K}$ is the number of iterations assigned. In this iterative technique of generating the regularized solution, K acts as the regularization parameter. It is very important to generate iterations up to an optimal number because the iterative solution can sometimes converge faster and if $\boldsymbol{K}$ is greater than K-optimal, the estimated solution might diverge from the true solution. The equation for the Kth iterative CG approach is given by:

$$X_{(K)} = X_{(K-1)} + \alpha_{(K)} \rho_{(K)}, \tag{4.17}$$

where, $\boldsymbol{X_K}$ is the Kth iterative approximation of $\boldsymbol{X}$ which is updated in each iteration by a multiple, $\boldsymbol{\alpha_K}$, of the search direction vector, $\boldsymbol{\rho_K}$. The conjugate gradient least squares equation is given by:

$$X_{reg.} = \min \left\| b - AX \right\|_2^2 . \tag{4.18}$$

Eq. (4.18) is similar to (4.12) – Tikhonov regularization technique, only difference is in eq. (4.18), $\lambda = 0$, nullifying the regularization term. Hence, in this technique the number of iterations, $\boldsymbol{K}$, acts as the regularization parameter.

Computational cost and storage requirements are certainly the prime factors in choosing a particular regularization technique for a specific application. Iterative methods for estimating a regularized solution of an ill-posed problem are fast gaining popularity due to their low computational cost and low storage requirements even for large-scale problems as compared to direct methods of regularization.

4.5 Simulation Results

In order to compare the full-solution of the IISR scheme with a wide class of super-resolution techniques, the problem of super-resolution is reformulated to Tikhonov regularized optimization and the iterative conjugate gradient method is adopted for minimizing the objective function, $\boldsymbol{\Phi}$ given in eq. (4.6). The conjugate gradient algorithm is an effective method of solving large-scale problems. The convergence rate for the method is quite rapid and, for most cases, the method is faster than the steepest descent, for example. For purely numerical and practical reasons it is convenient to remodel eq. (4.6) by absorbing the second term (regularization) into the first one, on the expense of increasing the size of matrixes involved in the calculations:

$$\Phi(X) = \left\| \begin{bmatrix} A \\ \lambda Q \end{bmatrix} X - \begin{bmatrix} b \\ 0 \end{bmatrix} \right\|^2 . \tag{4.19}$$

Several experiments of reconstructing high-resolution imagery using the optimization approach have been conducted. To have a uniform continuity and also to compare the reconstruction quality of both the IISR and optimization technique, super-resolution of the 'man' sequence [1024 x 1024] is considered again. Matrix $\boldsymbol{A}$ contains a Gaussian blur kernel [12 x 12] with a standard deviation of 2 pixels and a down-sampler to generate the artificial LR frames at a ratio of 16. It is assumed that the relative motion between the frames is well approximated by a single shift vector for each frame. Unlike many other papers investigating super-resolution reconstruction, the down-sampling ratios used in the simulation are large (usually 12, 16 or 20), which brings the calculations closer to reality. Note that the physical enhancement of the resolution achieved during the reconstruction process is usually smaller than the reconstruction magnification ratio. A discrete representation of Laplacian, fig. 4.1(a) was used as the regularization matrix $\boldsymbol{Q}$ with optimal values of the regularization parameter, λ, calculated using the L-curve criterion.

4.5.1 Noiseless LR Frames

The optimization procedure was applied to the same sequence of noiseless LR frames that were used in the simulation results of chapter 3. Again, for practical reasons, the number of LR frames used is kept as low as 10 and the optimal regularization parameter, λ, in this case is 0.1. Figure 4.4 shows one of the ten

Fig. 4.4 One of the 10 noiseless LR frames of the 'man' sequence

Fig. 4.5 High-resolution image reconstructed using the optimization procedure from only 10 noiseless LR frames of the 'man' sequence [1024 x 1024] with an E_{rmse} = 0.14959 and E_{psnr} = 64.633 dB

noiseless LR images of the 'man' sequence and fig. 4.5 shows the reconstructed high-resolution image.

The optimization procedure successfully generates a high-resolution image using only 10 noiseless LR frames. The RMSE calculated between the actual 'man' high-resolution image and the reconstructed one is 0.14959 and a PSNR value of 64.633 dB. Referring to fig. 3.11, the IISR-generated high-resolution image, the calculated RMSE was 0.14993 and PSNR = 64.6131 dB. By comparing the full solution of the optimization problem with the calculations based on the fast iterative-interpolation super-resolution (IISR) method (see fig. 4.6), one can notice the slight improvement over the IISR result as illustrated by the relatively small decline and rise in the RMSE and PSNR values respectively.

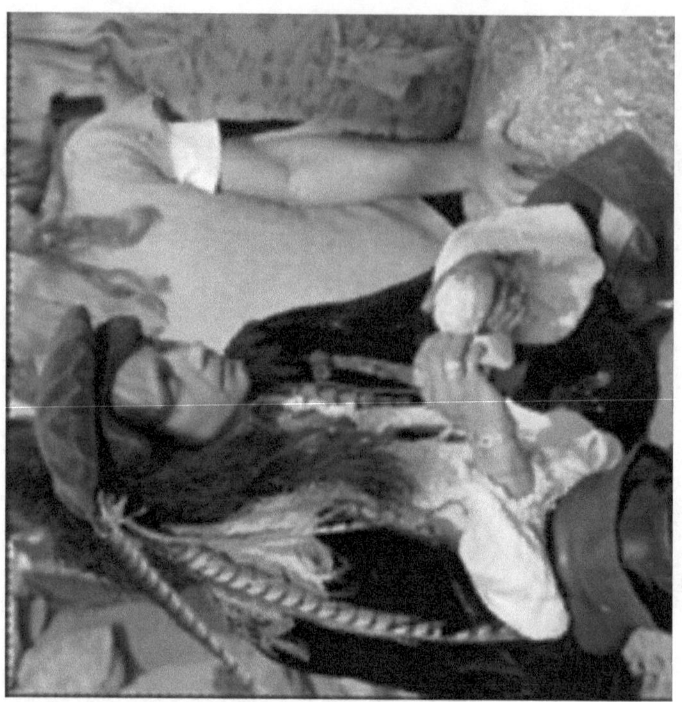

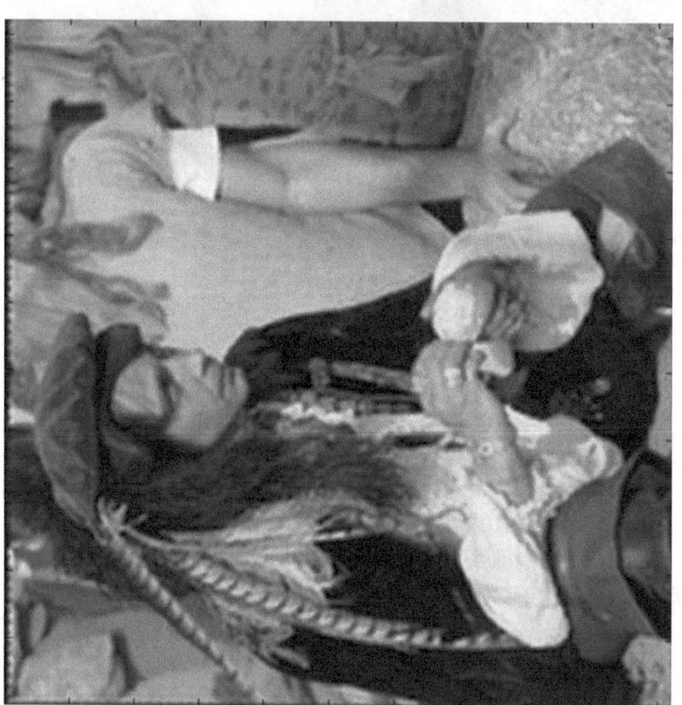

Fig. 4.6 IISR-generated high-resolution image (left-panel) and the high-resolution image generated using optimization approach (right-panel) of the 'man' sequence using only 10 noiseless LR frames

4.5.2 Noise Corrupted LR Frames

All the parameters are kept same as in the previous example except in this case a noise level of 20dB is introduced into the LR frames. Also, the optimal regularization parameter, λ, in this case is 0.065. Figure 4.7 shows one of the ten noise corrupted LR images of the 'man' sequence and fig. 4.8 shows the stabilized high-resolution image reconstructed by the optimization scheme.

Fig. 4.7 One of the 10 noise corrupted LR frames of the 'man' sequence

The RMSE calculated between the actual ‘man’ high-resolution image and the reconstructed one is 0.13837 and a PSNR value of 65.3097 dB. Referring to fig. 3.14, the IISR-generated high-resolution image, the calculated RMSE was 0.17301 and PSNR = 63.3696 dB. By comparing both these images (see fig. 4.9), it can be seen that the optimization scheme is able to achieve slightly better improvement over the IISR technique, as the regularization term is more powerful and suppresses a fair amount of noise in the final solution. It is easy to foresee that by using a larger number of LR frames, higher quality of reconstruction can be achieved. This can be seen in fig. 4.10.

Fig. 4.8 High-resolution image reconstructed using the optimization procedure from only 10 noise corrupted LR frames of the ‘man’ sequence [1024 x 1024] with an $E_{rmse} = 0.13837$ and $E_{psnr} = 65.3097$ dB

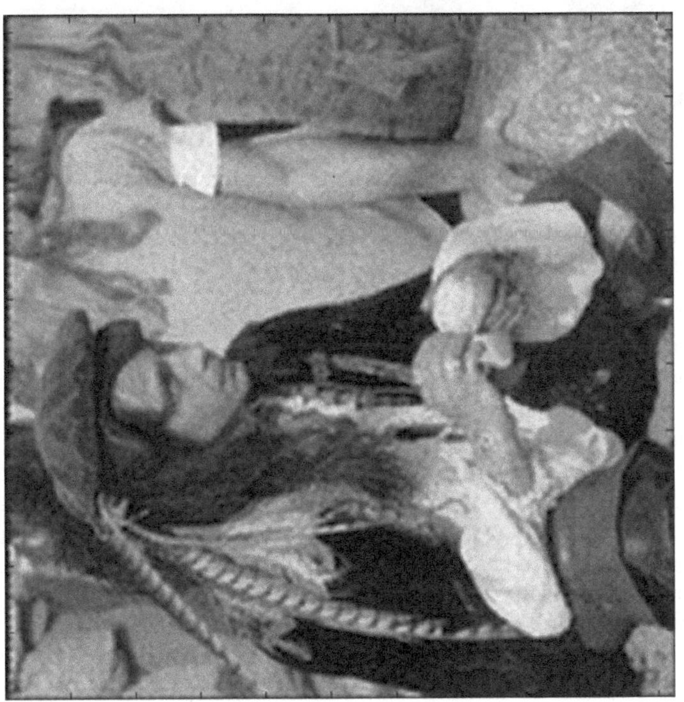

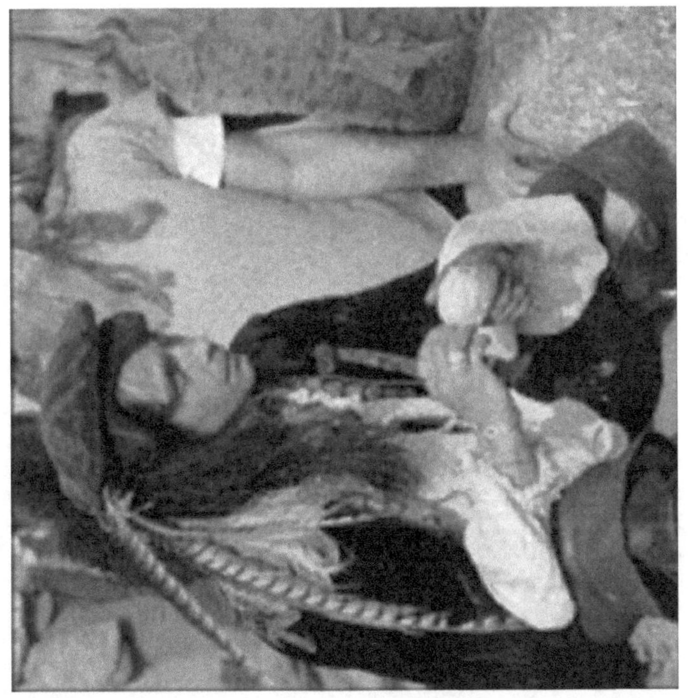

Fig. 4.9 IISR-generated high-resolution image (left-panel) and the high-resolution image generated using optimization approach (right-panel) of the 'man' sequence using only 10 noise corrupted LR frames

Fig. 4.10 High-resolution image reconstructed using the optimization procedure from only 40 noise corrupted LR frames of the 'man' sequence [1024 x 1024] with an E_{rmse} = 0.11914 and E_{psnr} = 66.6099 dB

4.6 Quantitative Analysis

To evaluate the reconstruction process of the optimization technique and also to compare it with IISR technique, quantitative analysis for the super-resolution of the 'man' sequence is presented in this section.

In the optimization technique, the regularization term plays a very important role of stabilizing the final solution. The amount of strength of this term is controlled by the regularization parameter, λ. In order to locate the optimum regularization parameter for a particular sequence of LR frames and to investigate the role of λ for the optimization process, computer simulations were executed for the minimization algorithm for a wide range of λ values. During calculations, the values of RMSE between the reconstructed image and the original test image were recorded, together with the values of the residual fitting error $\|b - AX\|^2$ and the stabilization term $\|QX\|^2$. These values were then used in construction of the L-curves and λ-curves for that sequence.

Examples of the resulting L-curve and λ-curve are shown in both, fig. 4.11 and 4.12 for noiseless and noise corrupted super-resolution reconstruction. As mentioned in the simulation results, the regularization matrix was chosen to be a discrete

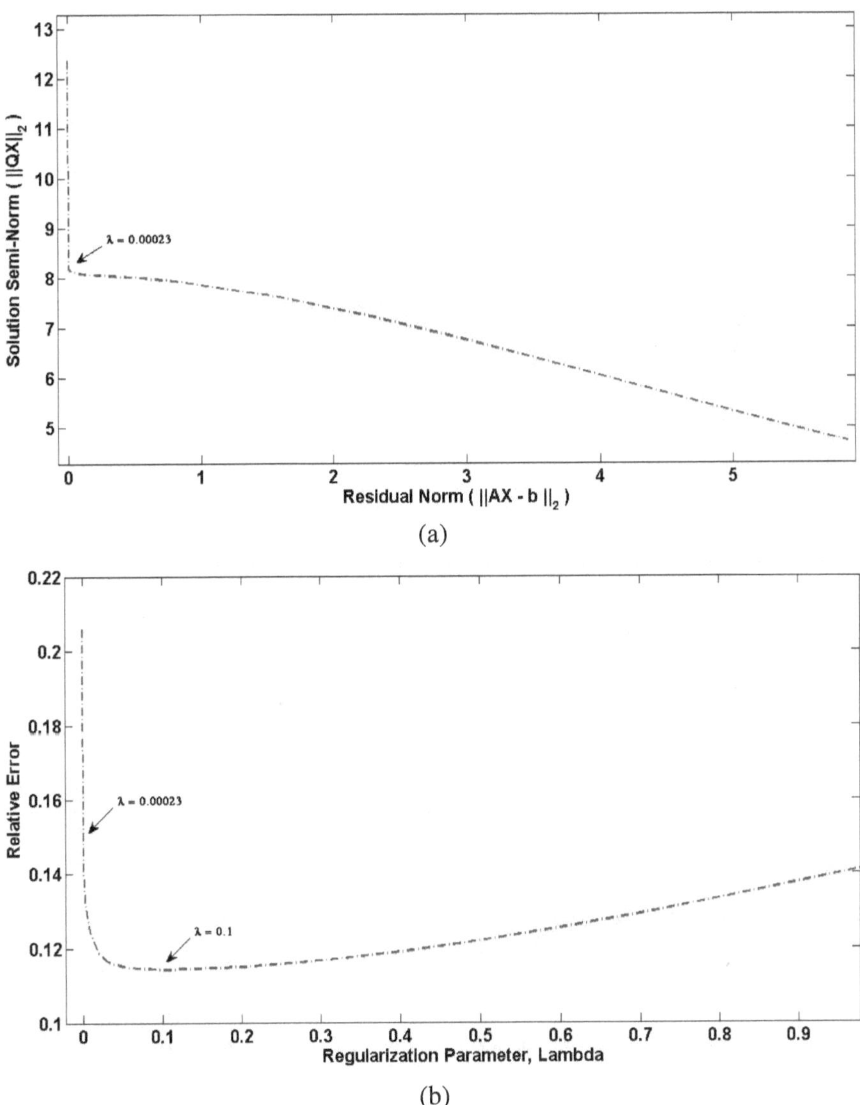

Fig. 4.11 (a) L-curve and (b) λ-curve for super-resolution reconstruction using only 10 noiseless LR frames of the 'man' sequence

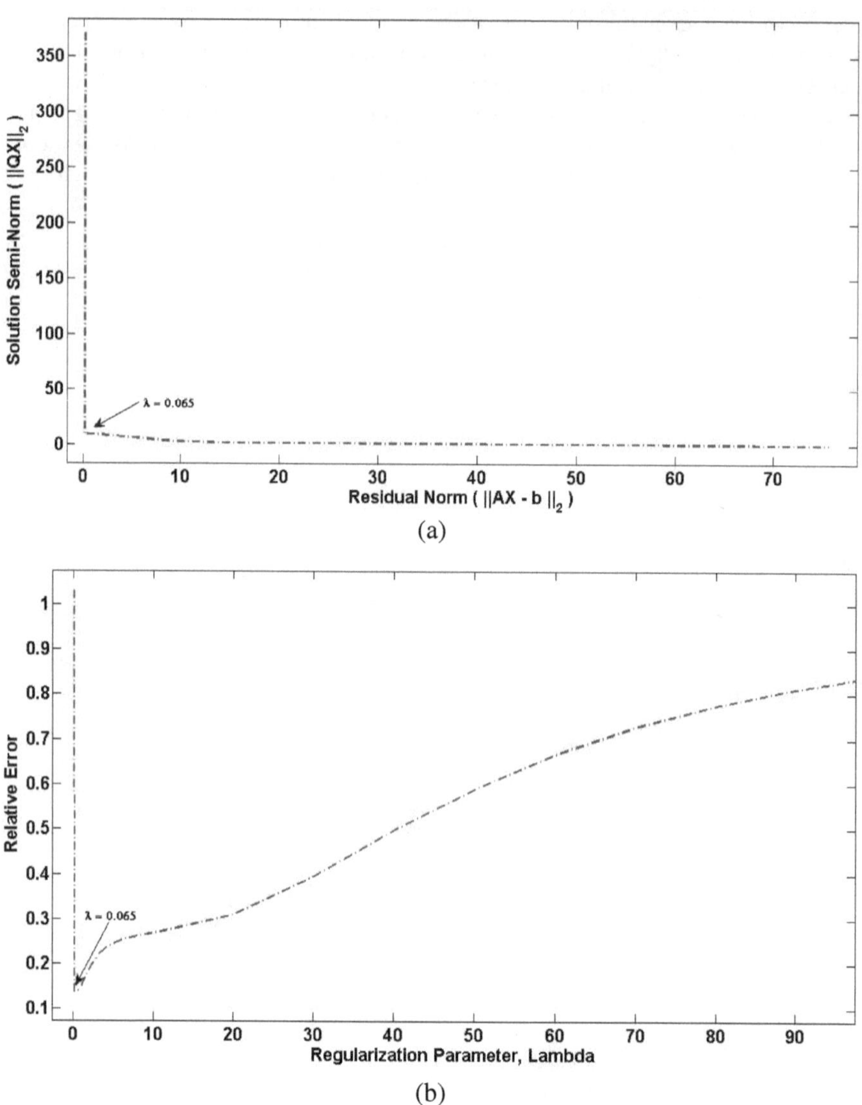

Fig. 4.12 (a) L-curve and (b) λ-curve for super-resolution reconstruction using only 10 noise corrupted LR frames of the 'man' sequence and the optimal $\lambda = 0.065$

representation of the Laplacian operator, as shown in fig. 4.1(a). Both curves have overall shapes that resemble the generic curves shown in, fig. 4.2 and 4.3.

In both the figures, 4.11(b) and 4.12(b), there is a gradual increase of the reconstruction error for λ larger than the optimal value, whereas for values of λ smaller than the optimal one, the reconstruction error grows very rapidly, which can be seen by the steep vertical graph. This behaviour has important practical implications: the super-resolution reconstruction is not very sensitive to the precise value of the regularization parameter when the parameter is larger than the true optimal value. This may explain why in many simulations good reconstructions were achieved with regularization parameter chosen in an *ad hoc* fashion. However, because the simulations were limited only to several examples of imagery, the observed insensitivity of super-resolution reconstruction to the exact value of λ may not be a universal property.

In fig. 4.11, the optimal value for λ as calculated by L-curve technique is 0.00023 whereas using the λ-curve technique, the lowest RMSE or best reconstruction quality is achieved at 0.1. Therefore, for the case of noiseless LR frames, the optimal value of λ is chosen to be 0.1 instead of 0.00023. On the other hand, in fig. 4.12, it is important to note that the optimal values for λ (0.065) as calculated from both methods are consistent with each other.

The rate of convergence of minimization routines strongly depends on the initial point that the optimization process starts with. To improve the convergence rate of optimization, the full-solution of the fast IISR technique is used as the starting point in the minimization process. It was observed that the final result of optimization did not significantly depend upon the initialization method. This was reassuring since the whole purpose of the regularization is to stabilize the solution. It was however, observed, that the speed of convergence was significantly improved. This is illustrated in fig. 4.13 which shows a comparison of convergence rates for the optimization procedure initiated by a blank image and by the final IISR reconstructed image using both, noiseless and noise corrupted LR frames.

The simulations illustrated in both curves of fig. 4.13 show that, for blank input image initialization, the optimization routine requires over 80 iterations to reduce the reconstruction error to the initial level of the final IISR reconstructed high-resolution image error. The further improvement of the reconstructed image is, however, rather small, reconfirming good accuracy of the proposed IISR reconstruction algorithm. It is also evident from the graphs, that this additional improvement over IISR result is computationally expensive since the convergence rate becomes slow.

Figure 4.14 summarizes the accuracy of super-resolution reconstruction as a function of number of LR frames utilized in the reconstruction process. It is clearly evident from the graphs that as the number of low-resolution frames utilized increases the efficiency of super-resolution reconstruction also improves. It is also clear from fig. 4.14(b) that even though the number of noise corrupted LR frames is increased, due to the robustness of the optimization routine, no noise amplification occurs (i.e.- noise is suppressed) causing further degradation of the reconstructed high-resolution image. This can also be seen by comparing the high-resolution images generated using noise corrupted 10 and 40 LR frames in fig. 4.8 and 4.10, respectively.

Computer simulations have also been performed with different forms of the regularization term in eq. (4.17). Four different forms of regularization matrix, $\boldsymbol{Q}$, have been implemented shown in fig. 4.1, which are based on various discrete representations of Laplacian and Biharmonic operators. Figures 4.15 and 4.16

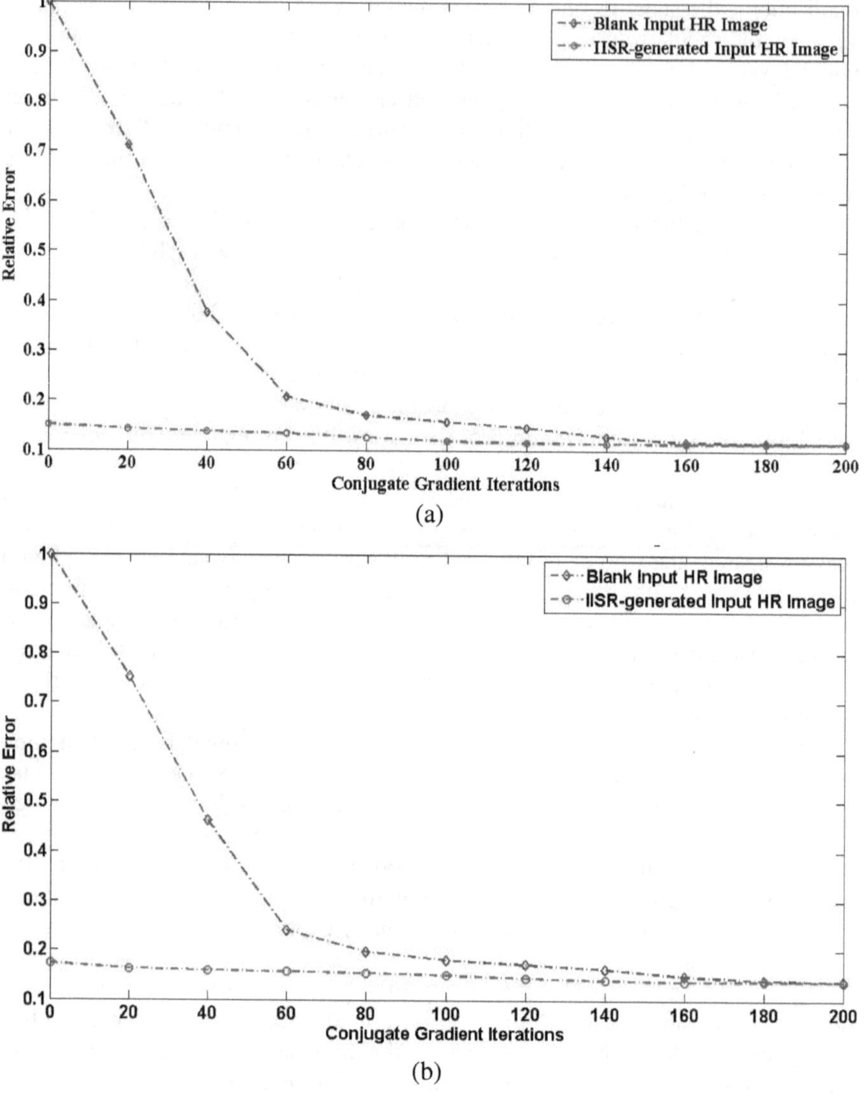

Fig. 4.13 Convergence rate for optimization super-resolution reconstruction of the 'man' sequence using only 10 (a) noiseless and (b) noise corrupted LR frames

illustrate the comparison of convergence rates of the optimization procedure for the 'man' sequence initiated by a blank image using both, noiseless ($\lambda = 0.1$) and noise ($\lambda = 0.065$) corrupted LR frames along with different regularization matrices, $\boldsymbol{Q}$, shown in fig. 4.1. Due to the extremely large size of $\boldsymbol{Q}$ for 24-neighbors, 2D-biharmonic operator, the computer simulations failed as MATLAB incurred an "out of memory" error message on a 1GB RAM, Intel Centrino (1.4 GHz) notebook. Therefore, in fig. 4.15 and 4.16, the convergence rate for the

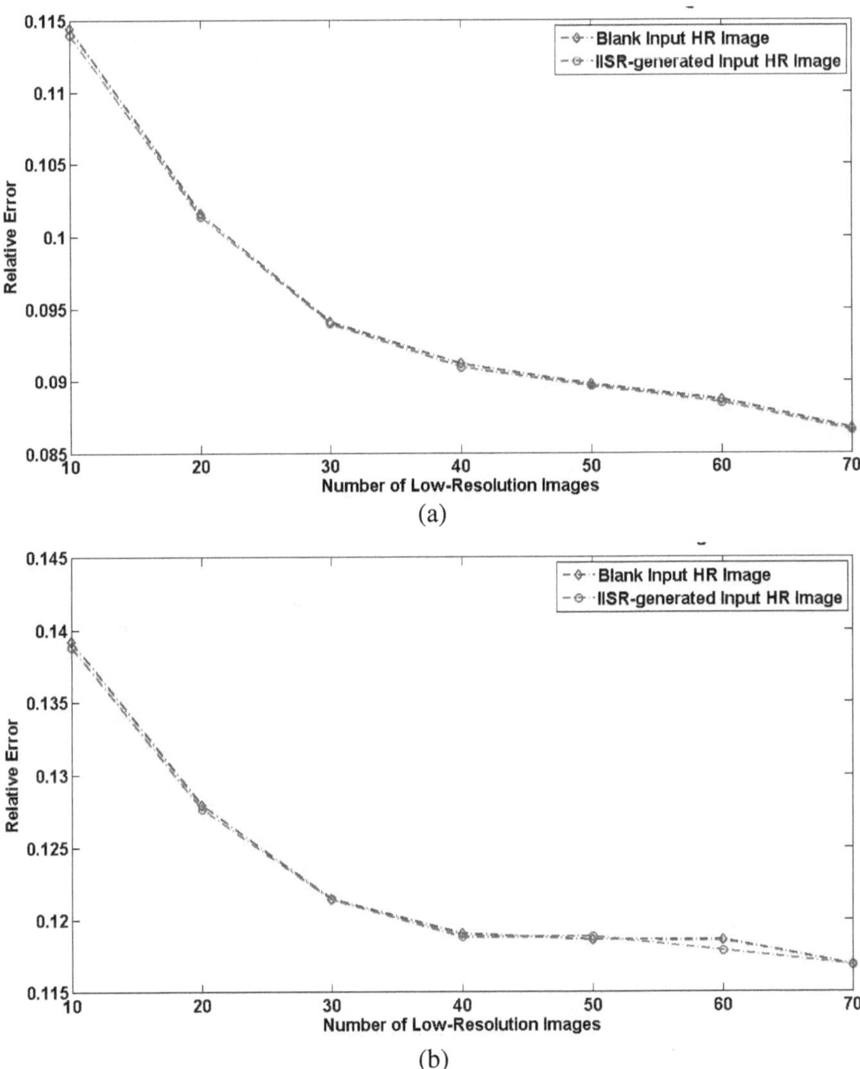

Fig. 4.14 Accuracy of super-resolution reconstruction of the 'man' sequence as a function of the number of (a) noiseless and (b) noise corrupted LR frames

'man' sequence using $\boldsymbol{Q}$ as the 24-neighbor, 2D-biharmonic operator has been omitted but for smaller high-resolution image sizes, for example a 512 x 512 high-resolution image, the data can be calculated. From the various computer simulations that have been performed for a smaller high-resolution image size, the tendency of the 24-neighbor regularization matrix is very similar to the 12-neighbor, 2D-biharmonic operator. In fig. 4.15(a) and 4.16(a) the 12-neighbor,

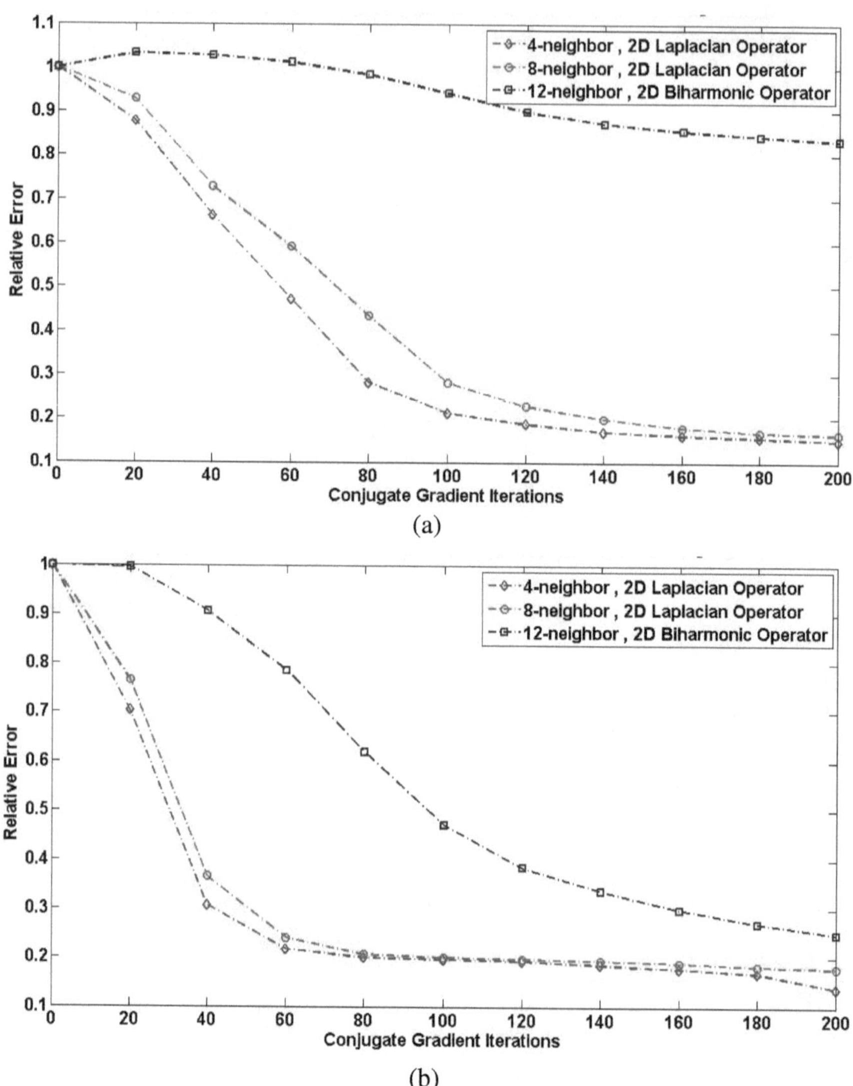

Fig. 4.15 Convergence rate for optimization super-resolution reconstruction of the 'man' sequence as a function of (a) 10 and (b) 40 noiseless LR frames using different regularization matrices (see legend), λ = 0.1 and a blank input HR image as the initialization

2D-biharmonic operator was observed to be less efficient as compared to the Laplacian operators for a lower number of LR frames used in the reconstruction process. On the other hand, with the increase in the number of LR frames used for the reconstruction process (see fig. 4.15(b) and 4.16(b), the biharmonic operator performs satisfactorily and shows the characteristic of strong noise suppression.

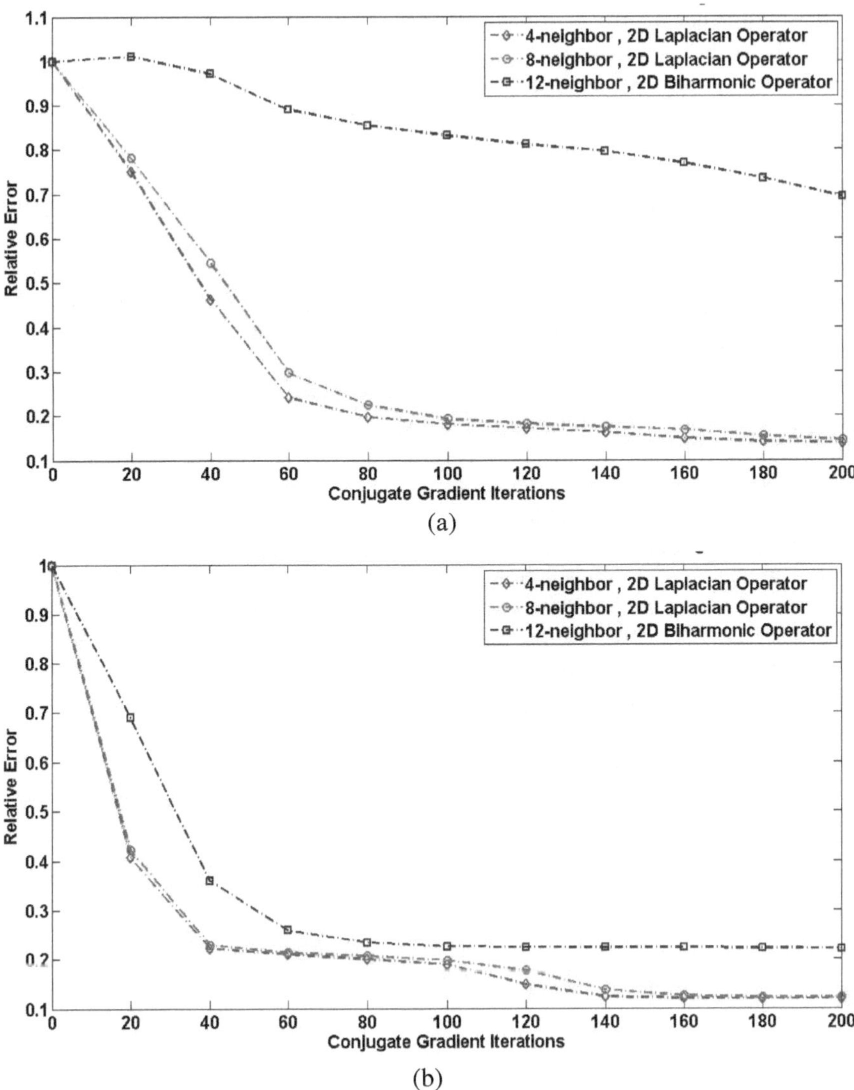

Fig. 4.16 Convergence rate for optimization super-resolution reconstruction of the 'man' sequence as a function of (a) 10 and (b) 40 noise corrupted LR frames using different regularization matrices (see legend), $\lambda = 0.065$ and a blank input HR image as the initialization

4.7 Summary

In this chapter, the optimization approach to image super-resolution is reinvestigated. The estimation problem is reformulated in terms of a Tikhonov regularized optimization problem and an iterative conjugate gradient technique is adopted for solving it.

To study the significance and influence of the regularization term over the accuracy of the reconstruction scheme in optimization, several different forms of regularization term were considered. The analytical data produced shows that the 2D Laplacian operators perform better than the 2D biharmonic operators especially when low numbers of LR frames are utilized for achieving super-resolution image reconstruction. The L-curve criterion is adopted for estimating the optimal value of the regularization parameter. It is also important to note that depending upon the image and other parameters such as the regularization matrix, the optimal value of λ changes. Therefore, it is critical that the L-curve graph be calculated every time the parameters are changed, to check for the optimal value of λ. This extra step ensures that the value of λ is neither too small nor too large. For simulated imagery, a new graphical analysis tool, λ-curve, is proposed. This graph can be utilized to confirm the optimal value of λ as calculated by the L-curve criterion. Unfortunately, in practice since the original scene is unknown, no relative error can be calculated between the original image and its attempted HR reconstruction. In such a scenario, λ-curve ceases to be of any help and one would have to depend completely upon the L-curve for providing an accurate regularization parameter, λ.

By comparing the full solution of the optimization problem with calculations based on the proposed fast iterative-interpolation super-resolution (IISR) method (see fig. 4.17), it is observed that the IISR reconstruction is reasonably accurate and that further improvement by the optimization procedure is relatively small and computationally expensive. The analysis also illustrate that the rate of convergence can certainly be accelerated by initializing the optimization procedure with the IISR solution instead of a blank input HR image.

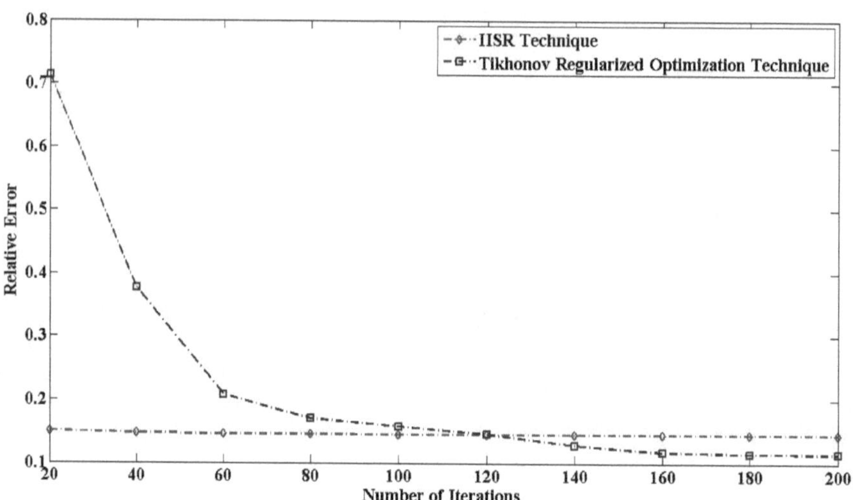

Fig. 4.17 Convergence plot for super-resolution image reconstruction of the 'man' sequence using the fast IISR and Tikhonov regularized optimization technique

Chapter 5
Image Registration for Super-Resolution

5.1 Introduction

Super-resolution reconstructs high-quality, high-resolution images by exploiting the fact that due to the relative motion between the camera sensor and the true scene, each aliased, under-sampled low-resolution frames acquired contains distinct incomplete and degraded scene information about the true scene. Therefore, in order for super-resolution to acquire this distinct scene information and successfully generate a high-resolution image, accurate knowledge of registration parameters is required for each of the input low-resolution frames. To formally define image registration, it is the process of geometrically aligning two or more images taken at different times, from different viewpoints and/or by different sensors. In various computer vision applications such as remote sensing, medical imaging, target detection, super-resolution imaging and many more, image registration is the most crucial intermediate process. To achieve accurate super-resolution image reconstruction, it is critical for image alignment to be precise. Mis-alignment of the under-sampled low-resolution frames will result in the reconstruction of an erroneous high-resolution image which may not be a true approximation of the original scene.

5.2 Methodology

As mentioned earlier image registration is an important prerequisite and is widely applied in many computer vision applications. A universal technique cannot be applied to solve all the tasks of registration in all the computer vision applications, due to the diverse nature of images involved, type of geometric warping, nature of transformation domain, and the required accuracy of registration. Even though numerous registration techniques have been proposed in the literature, the majority of them consist of the following steps for achieving image registration:

- *Control Point Detection and Matching*: Distinctive features such as boundaries, edges or corners are identified, manually or automatically, in both the target and the reference frame. These features are represented as control points in the images. A correspondence or matching is then established between the control points of the same scene in the target and the reference frame. The most difficult step of image registration is the accurate establishment of correspondence between a set of control points as it decides the accuracy level of the registration process.
- *Mapping function*: This task involves the computation of a geometric transformation model and its parameter estimation to accurately overlay the target frame over the reference frame.

V. Bannore: Iterative-Interpolation Super-Resolution Img. Recons., SCI 195, pp. 77 – 91.
springerlink.com

- *Image Resampling*: Utilizing the mapping function computed in the previous step, the target frame is resampled and aligned with the reference frame.

5.3 Image Registration Techniques

In the last two decades, extensive research has been carried out to develop new image registration techniques for solving real-world problems. One of the main concerns with any registration technique is the complexity and the high computational cost involved in the estimation of motion parameters.

An extensive survey of image registration techniques has been documented in [65]. The author not only describes merits and demerits of a broad range of registration techniques but also classified the techniques depending upon the variations in the images. The paper also extensively describes the theory of image registration. A more recent survey on the latest registration techniques deployed in various computer vision applications is documented in [66]. In this paper, the author has classified the techniques according to the four basic steps of image registration procedure: feature detection, feature matching, transform-model estimation, image re-sampling and transformation.

In [67], Zhao *et al.* discusses the effects of alignment and warping errors on super-resolution. They also introduced a new concept of optical flow consistency and flow accuracy. The error analysis provided in the paper indicates that in the case of small noise, optical flow is feasible for super-resolution. In [68], the authors provide a review of techniques for the computation of motion and structure from a sequence of monocular images. The paper gives a comparative view of the two most distinct approaches for the computation of motion - feature based and optical flow based techniques. The concept of optical flow was first introduced by Horn and Schunck in [69]. Due to the relative motion between the camera sensor and the objects in the scene, distinct information can be obtained by analyzing the difference between the time varying sequences of images. Optical flow is the process of calculating the instantaneous velocity of brightness patterns in the image. Optical flow relies on the assumption that any changes in spatial-temporal intensities in the image sequence are entirely due to the motion between the sensor and the scene (or object) [70]. With additional constraints, optical flow technique is used for motion estimation since it is possible to differentiate the pixels representing motion from the ones representing static environment. The technique also helps to analyse the direction of motion in the image.

In [71], the authors are primarily interested in generating high-resolution version of the input low-resolution video sequence of faces. The existing super-resolution approaches are not applicable for achieving super-resolution of video sequence of faces because faces are neither rigid nor planar and cause self-occlusion. They propose a super-resolution optical flow algorithm which not only generates the optical flow of the entire input video but also simultaneously produces the high-resolution version of the video. The idea of simultaneous estimation of registration parameters and high-resolution image is not new and was also proposed by Hardie *et al.* in [72]. They introduced a joint MAP framework for estimating the image motion parameters and generating the high-resolution image. The registration parameters are iteratively updated using the best existing high-resolution image estimate.

Pyramidal architectures are commonly employed in the registration process [5, 73-75]. Such architecture not only decreases the overall computational cost but also increases the accuracy of registration parameter estimation. Pyramid based registration techniques therefore are more desirable and accurate than the regular single-scaled based techniques. A number of ways have been documented in literature for constructing image pyramids but the most commonly utilized technique is the Gaussian and Laplacian pyramid structure proposed by Burt *et al.* in [76]. The image pyramid is a multi-resolution representation of an image constructed by successive levels of band-passed, sub-sampled images. A more general description of image pyramids in the field of image processing is given in [77]. Bergen *et al.* in [5] proposed a hierarchical framework for the computation of motion parameters. The authors have combined four motion models - affine flow, planar surface flow, rigid body flow and general optical flow into a single framework. A coarse-to-fine refinement strategy is implemented for computational efficiency. More iteration are performed on the coarsest level of the image resolution pyramid and once acceptable motion parameters are estimated on that level, they are then used as initial conditions on a more finer level of the resolution pyramid. This strategy enables the registration technique to converge rapidly and accurately. Ravi *et al.* in [74] used the Laplacian pyramid approach based on [5] for the registration of multisensor images such as infrared and radar. Thevenaz *et al.* in [75] proposed a full 3-D registration algorithm utilizing a cubic spline model for generating the multi-resolution image pyramid [78, 79]. Their algorithm is also applicable to a 2-D model, without any further modifications. The general affine transformation model in [75] can not only be restricted to a rigid body transformation but even to a simple translation model.

The imaging model considered in this study is given by equation (3.14). The matrix $\boldsymbol{A}$, is the imaging process consisting of 3 major processes, namely - warping, blur and down-sampling. So far in previous chapters it has been assumed that the motion parameters that constitute the warping matrix are known beforehand. It is important to note that among all the 3 processes of matrix $\boldsymbol{A}$, precise estimation of the warping matrix (accurate estimation of the relative motion between the sequence of low-resolution frames), is of the highest significance since a mis-registration can cause the super-resolution reconstruction process to construct a high-resolution frame of extremely poor quality. Even though the aim of this research is fast super-resolution image reconstruction, the issue of accurate motion estimation is also addressed.

5.4 Hierarchical Motion Estimation

As mentioned in the previous section, registration techniques based on the pyramidal architecture certainly prove to be more advantageous in terms of accuracy and computational efficiency. The hierarchical framework described in this section, for the computation of motion parameters, is adopted and implemented based on the work described by Bergen *et al.* in [5]. The basic components of hierarchical motion estimation are,

- Pyramid Construction.
- Motion Estimation.

- Image Warping.
- Coarse-to-fine Refinement.

The multi-resolution image pyramid is constructed using the Gaussian pyramid scheme described in [76]. The basic idea of Gaussian pyramid scheme is illustrated in fig. 5.1. The original image with fine resolution forms the base of the Gaussian pyramid, level 0. The resolution and sample density of the image at level 0 is reduced to half to form the level 1 image. In a similar way, the process continues to form the next higher level till the specified number of levels is reached forming the Gaussian pyramid structure with the tip of it being the coarsest resolution.

At each level the motion parameters that define the correspondence between the images are estimated by minimizing the objective or error function using the Gauss-Newton method. The error function, ***E***, is given as:

$$E(u,v) = \sum_{x}\sum_{y}[\, I_1(x,y)^l - I_2(x-u(x,y), y-v(x,y))^l\,]^2, \tag{5.1}$$

where, (x, y) denote the spatial coordinates at a particular position in images, I_1 and I_2; (u, v) represent the motion flow field between the images, the parameters of which need to be estimated by minimizing the error function and l denotes the level of image pyramid. Assuming a global motion model, affine transformation is used to approximate the relative motion flow field between the images. An affine model describes motions such as rotations, scaling, shearing and translation.

Thus, the affine model may be expressed as:

$$\begin{bmatrix} u \\ v \end{bmatrix} = A\begin{bmatrix} x \\ y \end{bmatrix} + \begin{bmatrix} a_x \\ a_y \end{bmatrix}. \tag{5.2}$$

In this research, the focus is on calculating parameters to describe only rotation, scale and translation,

$$A = \begin{bmatrix} a_{11} & -a_{12} \\ a_{21} & a_{22} \end{bmatrix}, \; a_{11} = a_{22} \; and \; a_{12} = a_{21}. \tag{5.3}$$

The motion of the entire region is then specified by vector ***a***,

$$a = \begin{bmatrix} a_{11} & a_{12} & a_{21} & a_{22} & a_x & a_y \end{bmatrix}^T. \tag{5.4}$$

The estimated motion parameters are then used to compute the motion flow field, given in eq. (5.5), which in turn is used for warping the target frame over the reference frame.

$$\begin{aligned} u(x,y) &= a_{11}x - a_{12}y + a_x \\ v(x,y) &= a_{21}x + a_{22}y + a_y \end{aligned}. \tag{5.5}$$

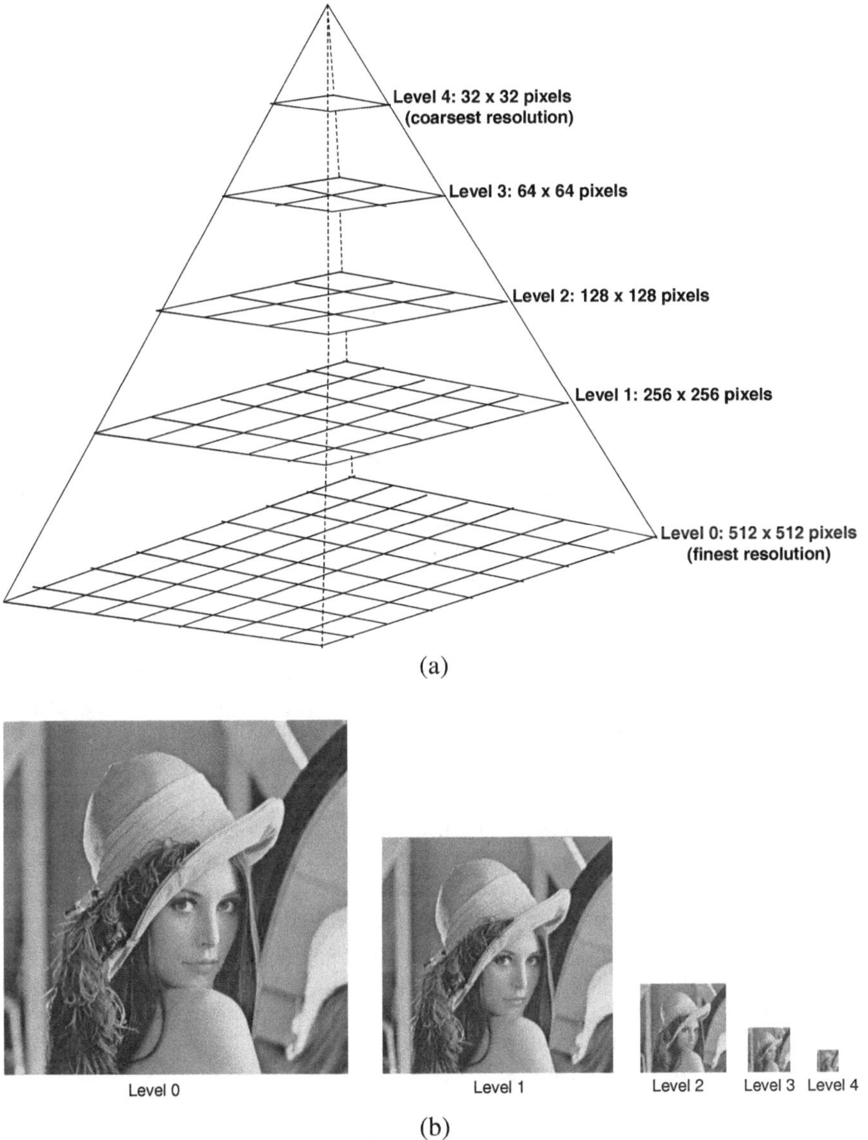

Fig. 5.1 (a) Gaussian image pyramid structure. (b) Five level Gaussian pyramid of 'Lena' image following the resolution structure shown in (a).

A coarse-to-fine refinement strategy is implemented for improving the overall performance of the registration technique. By adopting a hierarchical strategy, the search window can be limited because small displacements in finer resolutions represent large ones in coarser resolution and therefore the motion estimation becomes much more efficient. The motion parameters are estimated at each resolution starting from the top of the pyramid, which is the coarsest level. Once convergence is achieved at that level, the estimated motion parameters are then used as initial estimates for computing the parameters at the next level, down the pyramid. Therefore, at each new level the convergence is faster, since the transmitted motion parameters act as the initial conditions and are closer to the true parameters. The strategy also improves the computational efficiency, because more iteration are performed at the coarsest level of the pyramid where the amount of data is so greatly reduced that the computational load is negligible. Another attractive feature of using the pyramidal architecture is that it makes the registration technique robust to noise since the effect of noise at the coarsest level is almost indiscernible.

5.5 Simulation Results

Until now in the previous chapters, only translationally shifted low-resolution frames were considered for super-resolution image reconstruction. In this section, using the hierarchical motion estimation technique, rotationally and translationally shifted low-resolution frames are considered for creating a high-resolution frame.

Using hierarchical motion estimation technique discussed in the previous section, registration parameters are estimated from a sequence of low-resolution frames (Lena) shown in fig. 5.2. Since the low-resolution frames are artificially generated, the ground truth information about motion parameters is known. Therefore, in this case, it is convenient to calculate the accuracy of registration technique used, that is, comparing the motion parameters estimated to the ground truth. The registration technique was successfully able to estimate all the motion parameters accurately.

Using the estimated motion parameters, the IISR system generates a high-resolution frame from a sequence of aliased, under-sampled, rotationally and translationally shifted low-resolution frames. Figure 5.3 shows the final IISR generated high-resolution frame for the 'Lena' sequence of low-resolution frames.

Figure 5.4 shows another sequence of aliased, under-sampled low-resolution frames. These frames too are artificially generated and hence ground truth information about motion parameters is known for evaluating the accuracy of the registration technique implemented. The motion parameters are estimated accurately and are exactly same as the ground truth information. Figure 5.5 shows the final IISR generated high-resolution frame for the 'Concentric Circles' sequence of low-resolution frames.

Tables, 5.1(a) and 5.1(b), enumerate the accuracy of estimation of both, rotational and translation parameters, in a noiseless environment using the pyramidal registration technique. The statistical data presented in the tables was generated using all the 10 low-resolution frames of the 'Concentric Circles' sequence, shown in fig. 5.4. The ground truth data for both rotational and translational shifts has been provided for each of the low-resolution frames in the tables. Also, the total number of levels in the image pyramid, used for the motion estimation, are provided to illustrate the advantages of using a hierarchical architecture.

To test the robustness of motion estimation technique, white Gaussian noise level of 20dB is added to the sequence of under-sampled images of the

Fig. 5.2 Rotationally and translationally shifted noiseless low-resolution frames [85 x 85], of the 'Lena' sequence

Fig. 5.3 IISR-generated high-resolution image, [1020 x 1020] with RMSE = 0.068778 and PSNR = 71.3818 dB

'Concentric Circles' sequence shown in fig. 5.4 and are illustrated in fig. 5.6. Motion parameter estimation is then carried over these noise corrupted frames. The estimated parameters, both rotational and translational, for all the low-resolution images, using different number of levels of the image pyramid are populated in tables, 5.2(a) and 5.2(b). An example of the 4 - level multi-resolution image pyramid for one of the 10 noise-corrupted low-resolution frames is shown in fig. 5.7. Although the resolution is different at each level, the size has been expanded to the size of the base level of that pyramid.

Also, the effect of low-pass filtering is clearly evident from fig. 5.7. The effect of noise at the coarsest level, level 3 of fig. 5.7, is certainly indiscernible making it easier for the accurate estimation of motion parameters as the smallest displacement at the finest level becomes a large displacement at the coarsest level.

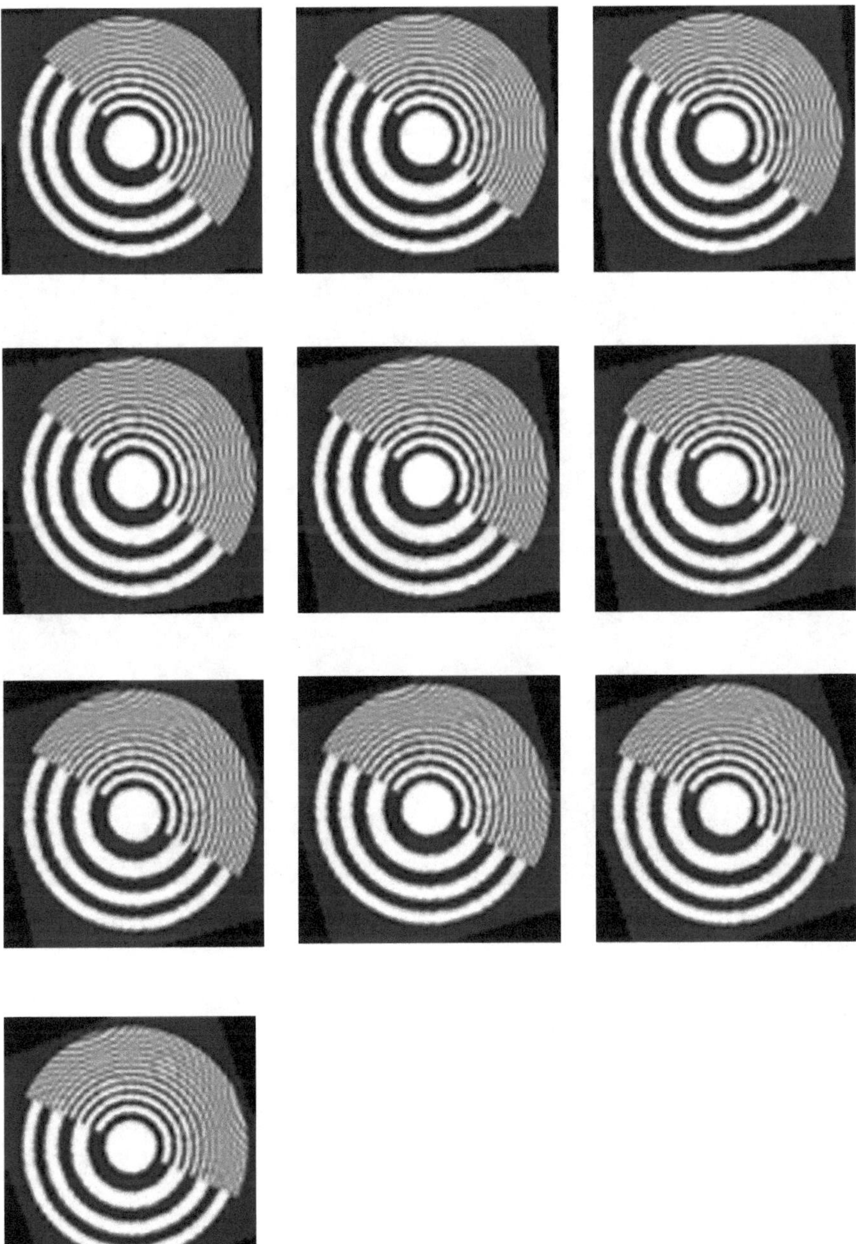

Fig. 5.4 Rotationally and translationally shifted noiseless low-resolution frames [85 x 85], of the 'Concentric Circles' sequence

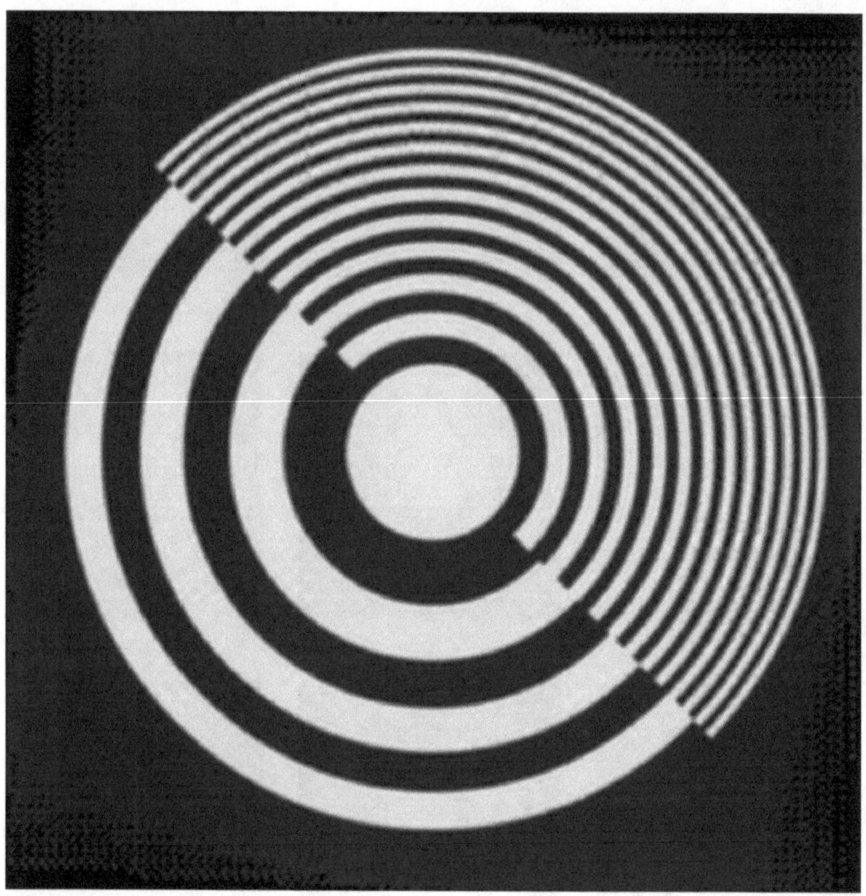

Fig. 5.5 IISR-generated high-resolution image, [1020 x 1020] with RMSE = 0.051681 and PSNR = 73.8641 dB

Table 5.1(a) Estimation of rotational shifts in noiseless low-resolution frames of the 'Concentric Circles' sequence

Low-Resolution Frames (θ in degrees)	Number of Levels in Image Pyramid			
	1	2	3	4
LR1 (θ = 2)	1.9453	1.9672	1.9672	1.9672
LR2 (θ = 4)	4.0495	4.0542	4.0542	4.0542
LR3 (θ = 6)	5.9539	5.9913	5.9913	5.9913
LR4 (θ = 8)	7.9445	7.9439	7.9438	7.9438
LR5 (θ = 10)	9.8172	9.9451	9.9453	9.9453
LR6 (θ = 12)	11.8329	11.9256	11.9256	11.9255
LR7 (θ = 14)	13.8318	14.2057	14.2059	14.2058
LR8 (θ = 16)	16.0731	16.1356	16.1351	16.1313
LR9 (θ = 18)	16.9218	18.2192	18.2217	18.2111
LR10 (θ = 20)	18.4856	20.0581	20.0715	-

Table 5.1(b) Estimation of translational shifts in noiseless low-resolution frames of the 'Concentric Circles' sequence

Low-Resolution Frames (dx,dy)	Number of Levels in Image Pyramid			
	1	2	3	4
LR1 (9,5)	(9.031 , 4.928)	(9.131 , 4.724)	(9.031 , 4.928)	(9.131 , 4.724)
LR2 (4,11)	(4.001 , 10.867)	(4.671 , 10.764)	(4.001 , 10.867)	(4.671 , 10.764)
LR3 (8,9)	(8.013 , 8.769)	(8.736 , 8.165)	(8.013 , 8.769)	(8.736 , 8.165)
LR4 (0,1)	(0.0199 , 1.109)	(0.0182 , 1.545)	(0.019 , 1.109)	(0.018 , 1.545)
LR5 (5,2)	(5.041 , 2.114)	(5.139 , 1.766)	(5.041 , 2.114)	(5.139 , 1.766)
LR6 (8,0)	(8.016 , 0.0530)	(7.644 , 1.116)	(8.016 , 0.053)	(7.644 , 1.116)
LR7 (1,5)	(0.992 , 4.963)	(1.996 , 5.314)	(0.992 , 4.963)	(1.996 , 5.314)
LR8 (0,10)	(0.027 , 9.771)	(2.610 , 10.223)	(0.027 , 9.771)	(2.610 , 10.223)
LR9 (4,6)	(4.054 , 5.844)	(5.234 , 5.382)	(4.054 , 5.844)	(5.234 , 5.382)
LR10 (3,3)	(3.087 , 3.106)	(3.584 , 3.068)	(3.087 , 3.106)	(3.584 , 3.068)

Table 5.2(a) Estimation of rotational shifts in noise-corrupted low-resolution frames of the 'Concentric Circles' sequence

Low-Resolution Frames (θ in degrees)	Number of Levels in Image Pyramid			
	1	2	3	4
LR1 ($\theta = 2$)	1.934	2.019	2.019	2.0191
LR2 ($\theta = 4$)	3.9736	4.0131	4.0132	4.0132
LR3 ($\theta = 6$)	5.7355	5.8825	5.8827	5.8829
LR4 ($\theta = 8$)	8.007	8.0197	8.0197	8.0197
LR5 ($\theta = 10$)	9.6605	10.0552	10.0554	10.0554
LR6 ($\theta = 12$)	11.4308	11.8579	11.8579	11.8579
LR7 ($\theta = 14$)	12.9267	14.1968	14.1974	14.1975
LR8 ($\theta = 16$)	15.3704	16.1807	16.1801	16.1758
LR9 ($\theta = 18$)	15.3854	18.325	18.3265	18.3193
LR10 ($\theta = 20$)	15.9934	20.1116	20.1166	-

Table 5.2(b) Estimation of translational shifts in noise-corrupted low-resolution frames of the 'Concentric Circles' sequence

Low-Resolution Frames (dx,dy)	Number of Levels in Image Pyramid			
	1	2	3	4
LR1 (9,5)	(9.045 , 4.877)	(9.045 , 4.877)	(9.045 , 4.877)	(9.045 , 4.877)
LR2 (4,11)	(3.998 , 10.885)	(3.998 , 10.885)	(3.998 , 10.885)	(3.998 , 10.885)
LR3 (8,9)	(8.019 , 8.760)	(8.019 , 8.760)	(8.019 , 8.760)	(8.019 , 8.760)
LR4 (0,1)	(0.036 , 1.081)	(0.036 , 1.081)	(0.036 , 1.081)	(0.036 , 1.081)
LR5 (5,2)	(4.965 , 2.042)	(4.965 , 2.042)	(4.965 , 2.042)	(4.965 , 2.042)
LR6 (8,0)	(7.980 , 0.0428)	(7.980 , 0.042)	(7.980 , 0.042)	(7.980 , 0.042)
LR7 (1,5)	(0.975 , 4.924)	(0.975 , 4.924)	(0.975 , 4.924)	(0.975 , 4.924)
LR8 (0,10)	(0.056 , 9.779)	(0.056 , 9.779)	(0.056 , 9.779)	(0.056 , 9.779)
LR9 (4,6)	(3.942 , 5.802)	(3.942 , 5.802)	(3.942 , 5.802)	(3.942 , 5.802)
LR10 (3,3)	(2.974 , 3.067)	(2.974 , 3.067)	(2.974 , 3.067)	(2.974 , 3.067)

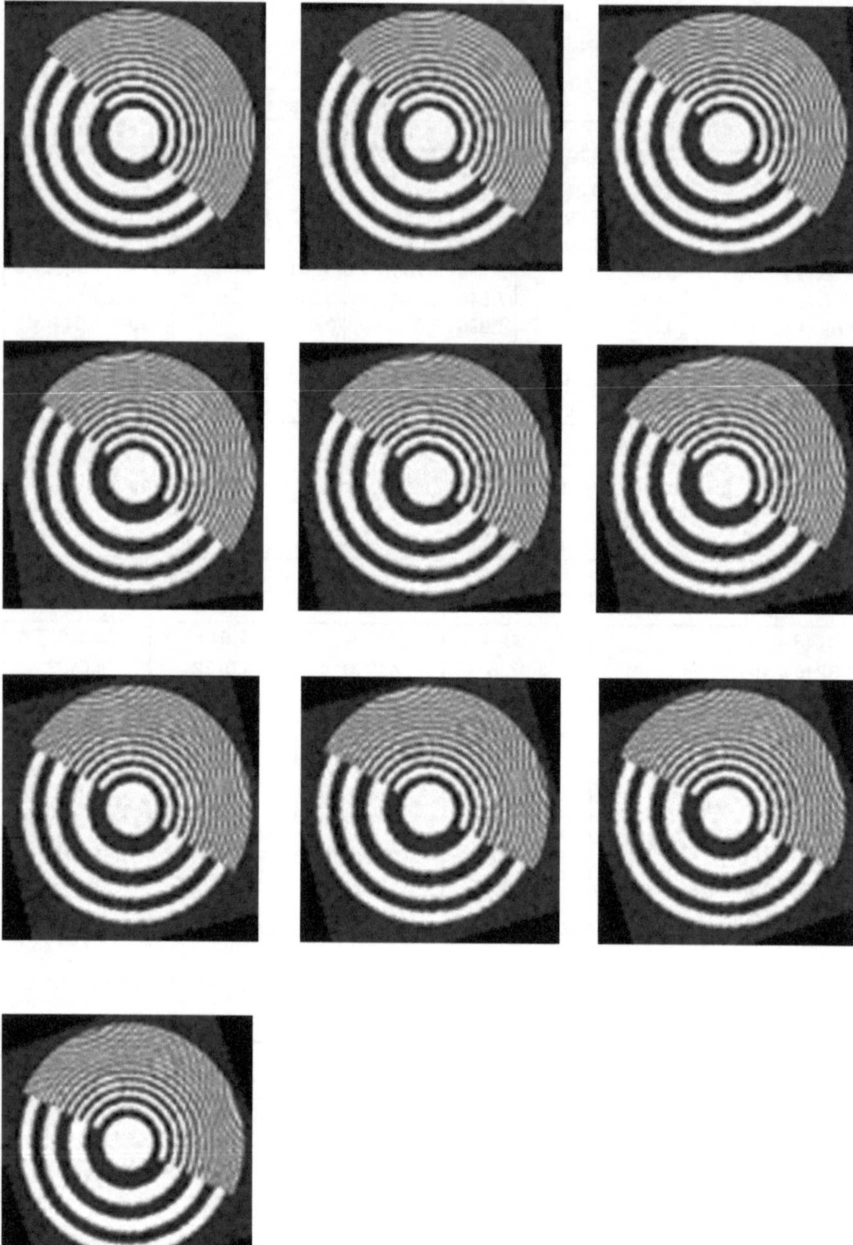

Fig. 5.6 Rotationally and translationally shifted, noise-corrupted low-resolution frames [85 x 85], of the 'Concentric Circles' sequence

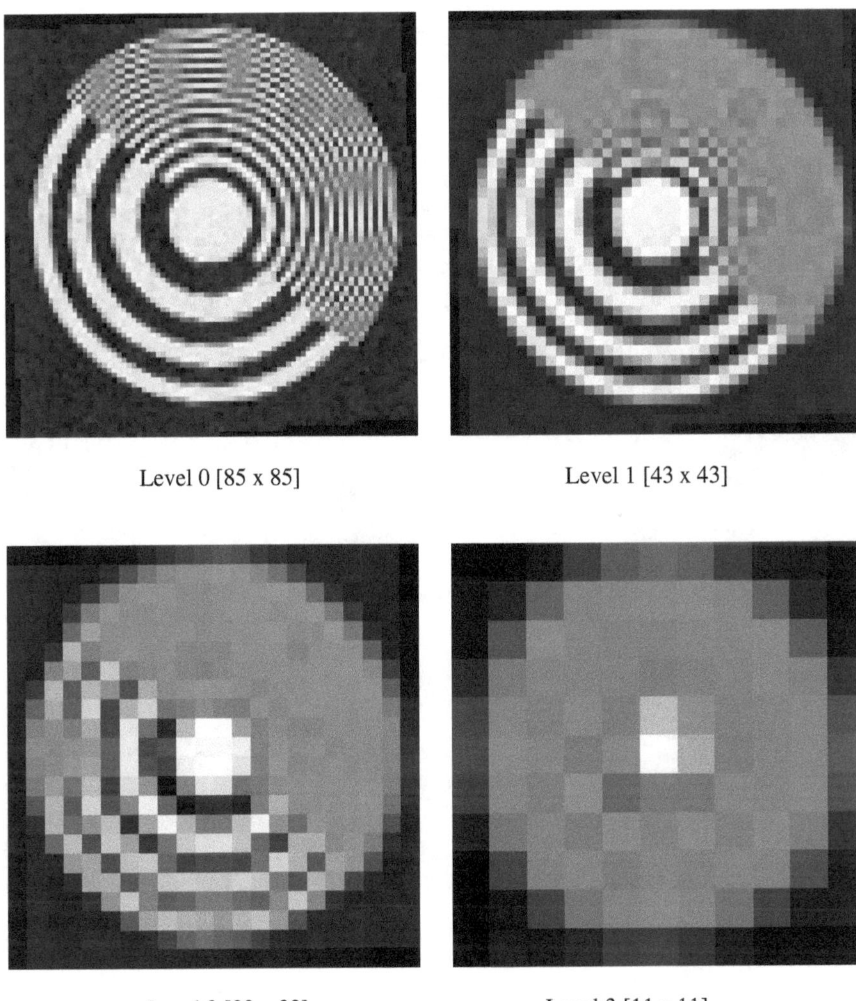

Fig. 5.7 Different levels of a 4 - level image pyramid with each level, expanded to the same size of image at level 0

The high-resolution image reconstructed, using the noise-corrupted low-resolution frames, is shown in fig. 5.8. One can clearly see from tables 5.1 and 5.2 that the pyramidal architecture certainly proves to be efficient in generating accurate estimations of motion parameters. The technique is also able to successfully estimate the motion parameters in a noisy environment. As evident from the tables, the technique is also able to accurately estimate the angle of rotation, as large as 20°.

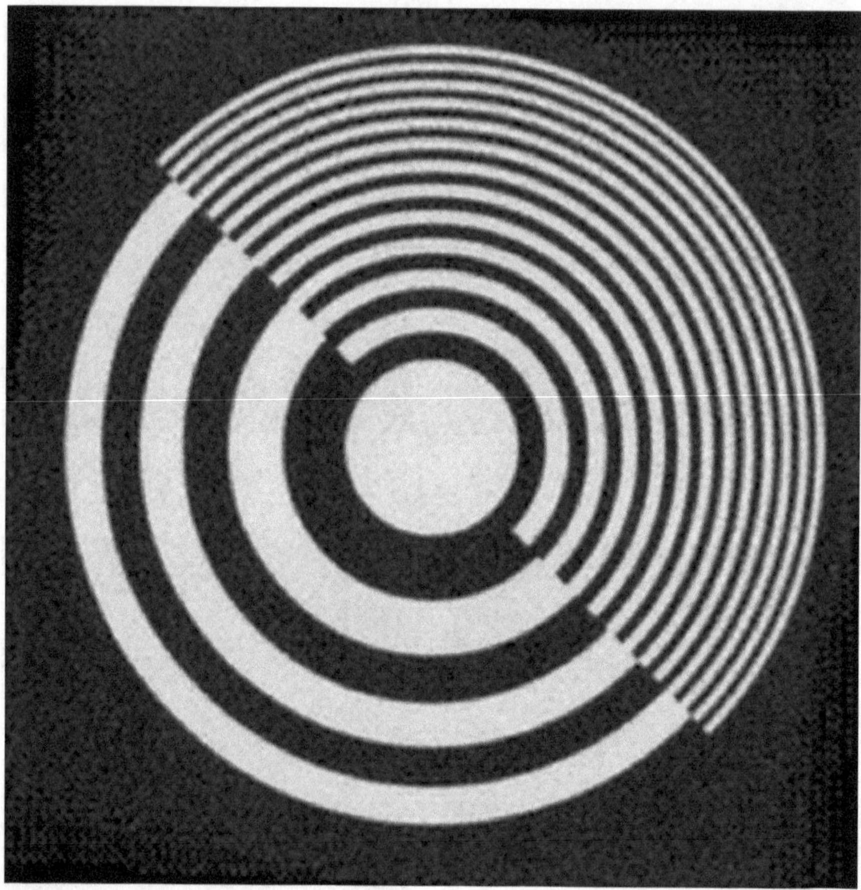

Fig. 5.8 IISR-generated high-resolution image, [1020 x 1020] with RMSE = 0.092203 and PSNR = 68.8359 dB

5.6 Summary

To achieve accurate reconstruction of high-resolution images from a sequence of under-sampled low-resolution frames, it cannot be emphasized more, how critical it is to align all the frames accurately. Almost all computer vision applications employ some form of image registration to acquire detailed information from the registered images.

Various image registration techniques have been introduced in the literature since last two decades. Among these, the hierarchical-based architecture for image registration proves to be an efficient and accurate technique for motion estimation. To simulate a realistic scenario for super-resolution, it is important to consider a general motion model for registering the low-resolution images. Therefore, based upon [5], the motion parameters are estimated from a sequence of rotated and translated, aliased, under-sampled, low-resolution frames.

The simulations provided in this chapter show that the multi-resolution image pyramidal architecture along with coarse-to-fine refinement strategy generates accurate estimation of motion parameters. In the case of noise contaminated low-resolution frames, employing such motion estimation scheme ensures that even with high noise, angle of rotation and translational shifts can be accurately recovered for precise alignment for maximum recovery of scene information to generate a high-resolution approximation of the true scene. It is also evident from the statistical data shown in tables, 5.1 and 5.2 that the estimated motion parameters are in complete agreement with the true parameters. The scheme converges for angle of rotation as large as 20°.

Being a hierarchical-based image registration technique, the method has a low computational load and is certainly efficient. The technique proves to be well-suited for incorporating it with the proposed fast IISR scheme, to provide a complete super-resolution image reconstruction package.

Chapter 6
Software Framework

6.1 Introduction

In this chapter, the software frameworks that have been developed during the research are presented:

- Iterative-Interpolation Super-Resolution (IISR).
- Image Super-Resolution Optimization (ISRO).

The techniques described in previous chapters have all been included in the above two mentioned softwares. For a better understanding, principle block diagrams for each of the above softwares will be presented. Each software package comprises of several routines which will be illustrated as modules in the principle block diagram of that software package and a description or functionality for each of these modules will be provided.

6.2 Iterative-Interpolation Super-Resolution (IISR)

This software package is based upon the novel approach presented in chapter 3 of this study. The aim of this software is to generate a high-quality, high-resolution image from a sequence of geometrically warped, aliased, and under-sampled low-resolution frames.

6.2.1 Block Diagram

The principle block diagram for the IISR system is shown in fig. 6.1. The imaging process acquires a sequence of under-sampled, low-resolution frames of the true high-resolution scene. The low-resolution frames undergo registration so as to estimate the motion parameters. Using these motion parameters, each pixel from each of the low-resolution frames is mapped on to a composite high-resolution grid image. The reconstruction process, then acts over this grid image to generate the first approximation of the true scene.

V. Bannore: Iterative-Interpolation Super-Resolution Img. Recons., SCI 195, pp. 93–103.
springerlink.com

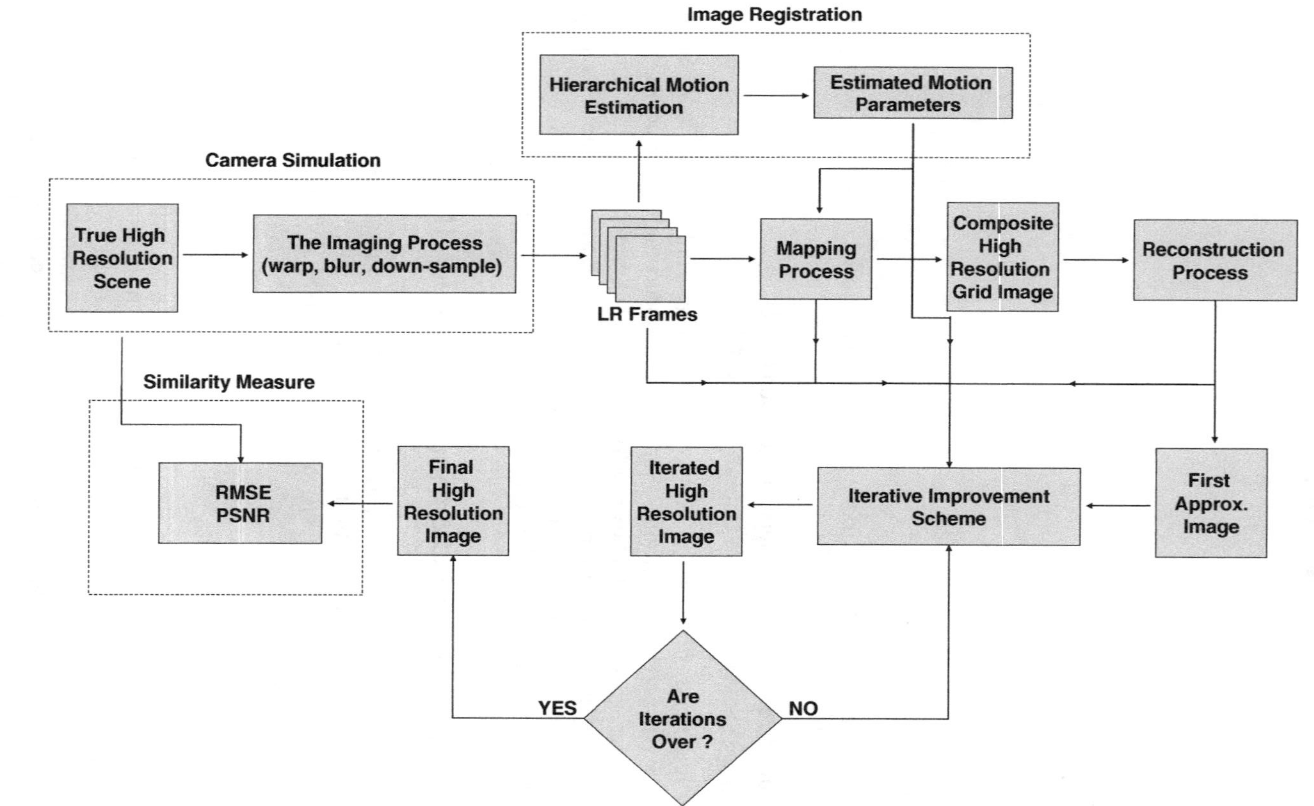

Fig. 6.1 Iterative-Interpolation Super-Resolution (IISR) Block Diagram

Depending upon the number of iterations specified, the first approximation undergoes iterative improvement to generate the final high-resolution estimate of the true scene.

6.2.2 Modules

As mentioned earlier, there are several routines that build up the IISR system and as shown in fig. 6.1, the software is completely modular to achieve maximum flexibility. The functionality of each of these routines or modules is explained below.

- *Camera Simulation*: This module is an approximate camera simulation routine. The imaging process introduces a desired amount of warping, blurring and down-sampling over the area of a high-resolution scene that is specified to be acquired, to generate an artificial low-resolution frame of that scene. The number of low-resolution frames to be generated can be specified in the imaging process.
- *Image Registration*: The aim of this module is to calculate the relative motion between two or more images. Using hierarchical motion estimation technique over the low-resolution frames, motion parameters are calculated.
- *Mapping Process*: Using the estimated motion parameters, this routine maps and up-samples each and every pixel from all the low-resolution frames onto a composite high-resolution grid. See fig. 6.2.
- *Reconstruction Process*: The reconstruction process interpolates the composite high-resolution grid image to generate the *first approximation* of the true high-resolution scene. Several interpolation kernels have been incorporated in the reconstruction process.

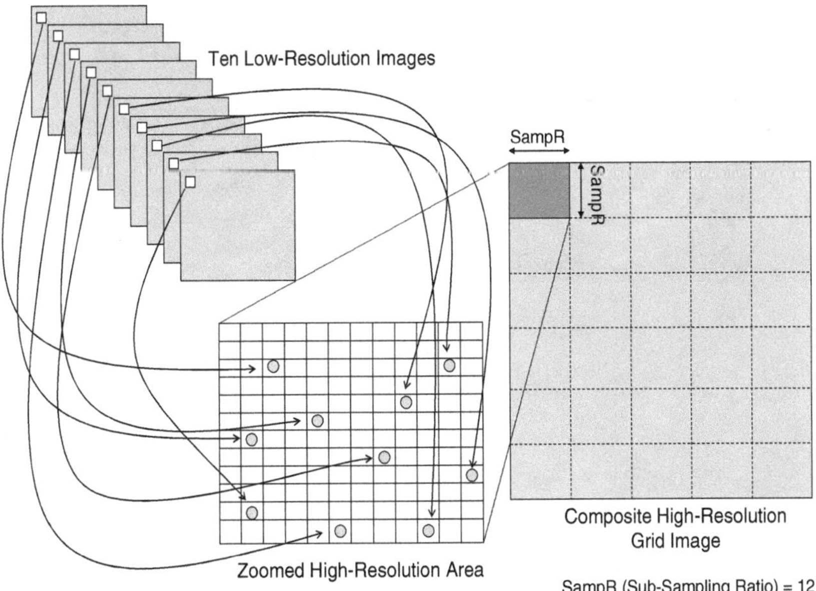

Fig. 6.2 Mapping Process

- *Iterative Improvement*: In this module, all processes mentioned above, that is, the imaging process, mapping and reconstruction process are incorporated. This module iteratively improves the first approximation until the specified number of iterations is finished and a final high-resolution estimate of the true scene is generated.
- *Similarity Measure:* To evaluate the reconstruction quality of the final IISR-generated high-resolution image, one could measure the similarity between the actual true scene or ground truth and the reconstructed scene. This would include calculating RMSE and PSNR. This module is only applicable when the original true high-resolution scene is known and available for comparison.

6.2.3 Software Tools

The IISR system is a MATLAB-based software package, built on MATLAB 7.1.0.246 (R14) Service Pack 3 using Image Processing Toolbox v5.1. MATLAB provides flexibility in designing and has a huge mathematical functionality for building various algorithms. The image processing toolbox provides an extensive suite of robust digital image processing and analytical functions. MATLAB's own graphical user interface development environment, GUIDE, was used to create graphical user interfaces (GUIs) to ease the use of the IISR system.

6.2.4 GUI Screenshot

Following is the screenshot of the IISR-GUI system.

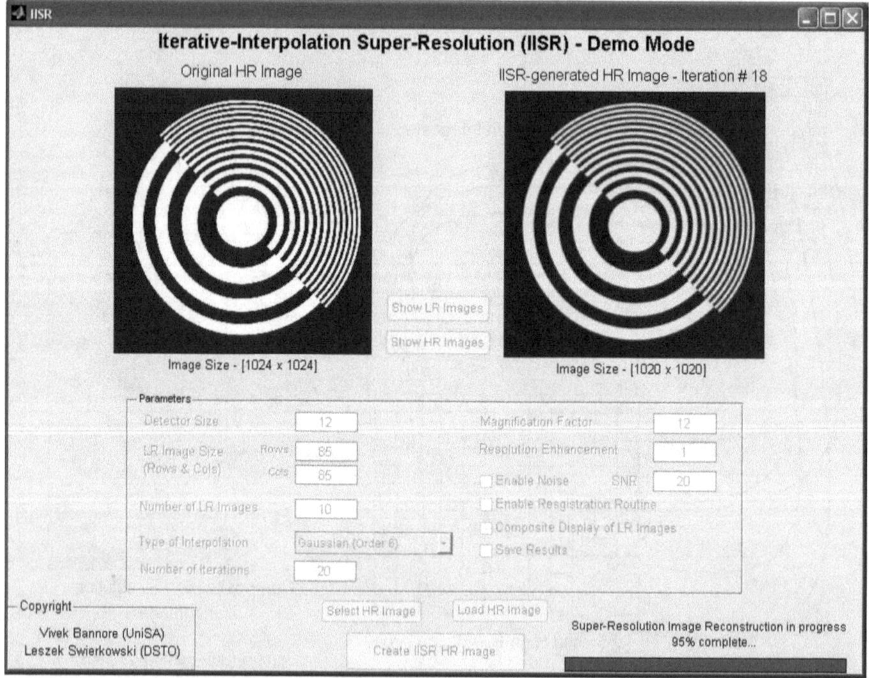

Fig. 6.3 IISR Graphical User Interface

6.2.5 Salient Features

Following are some of the most important features of the IISR software:

- The software can be executed in either a demo mode or work mode.
- In demo mode,
  - The user may choose a true high-resolution scene from which a sequence of artificially generated low-resolution images will be created and then using this sequence as the starting point, the IISR system will reconstruct the final estimate of the high-resolution frame.
  - The user can change the imaging process parameters, such as the decimation ratio, to obtain a different sequence of artificially generated low-resolution frames.
  - The user has the option of enabling or disabling the inbuilt registration routine. If disabled, the motion parameters used by the camera simulation module to generate the low-resolution frames will be used for further processing.
  - Overall, in the demo mode, the user only has a partial control over the IISR system.
- In work mode,
  - The user can provide a sequence of low-resolution frames in a desired format, for IISR to reconstruct the final estimate of the high-resolution frame.
  - The user also has the option of enabling or disabling the inbuilt registration routine. If disabled, the motion parameters need to be calculated independently of the system and provided to the IISR scheme for super-resolution image reconstruction to proceed.
  - Overall, in the work mode, the user has a full control over the IISR system.
- The user has the full ability to save the low-resolution and high-resolution frames as *.MAT/*.AVI files.
- The user can also opt to save the statistical data generated by the IISR system as a text file.

6.3 Image Super-Resolution Optimization (ISRO)

This software package was mainly developed in order to compare the full solution of the IISR-generated high-resolution image with the one generated by using the optimization approach. The software also helps in understanding the role of regularization in image super-resolution. The problem of estimating the high-resolution image is reformulated in terms of a regularized optimization procedure and the minimum of the resulting objective function is sought:

$$\min \left\| \begin{bmatrix} A \\ \lambda Q \end{bmatrix} X - \begin{bmatrix} b \\ 0 \end{bmatrix} \right\|^2 . \tag{6.1}$$

For more details, please refer to chapter 4.

6.3.1 Block Diagram

The principle block diagram for the ISRO system is shown in fig. 6.4. The imaging process acquires a sequence of geometrically warped, under-sampled, low-resolution frames of the true high-resolution scene. The low-resolution frames undergo registration so as to estimate the motion parameters. These parameters are then used to construct a large sparse sampling matrix. An initial estimate of the true scene along with regularization parameters, sampling matrix and the low-resolution frames are provided as an input to the minimization routine, to generate the final high-resolution estimate of the original scene.

6.3.2 Modules

The software is completely modular to achieve maximum flexibility. Functionality of each of these routines or modules is explained below.

- *Camera Simulation*: This module is an approximate camera simulation routine. The imaging process introduces a desired amount of warping, blurring and down-sampling over the area of a high-resolution scene that is specified to be acquired, to generate an artificial low-resolution frame of that scene. The number of low-resolution frames to be generated can be specified in the imaging process.
- *Image Registration*: The aim of this module is to calculate the relative motion between two or more images. Using hierarchical motion estimation technique over the low-resolution frames, motion parameters are calculated.
- *Sampling Matrix*: This process creates a sampling matrix which reverses the imaging process. The functionality of this matrix is similar to the mapping process of the IISR system.
- *Regularization Parameters*: The parameters of the regularization term shown in eq. (4.7), λ, the regularization parameter and $\boldsymbol{Q}$, the stabilization matrix are formed in this process.
- *Initialization*: In order to increase the convergence rate, the minimization algorithm can be initialized with some approximate estimate of the scene which needs to be super-resolved. This estimate could be a blank image too.
- *Minimization Routine*: An iterative conjugate gradient like method is implemented for finding the minimum of eq. (6.1). After completing a specified number of iterations, a final high-resolution estimate of the true scene is generated. The iterative techniques as compared to a direct method utilize matrix-vector multiplications thereby comparatively decreasing the computational time and storage requirements for huge sparse matrices and vectors.

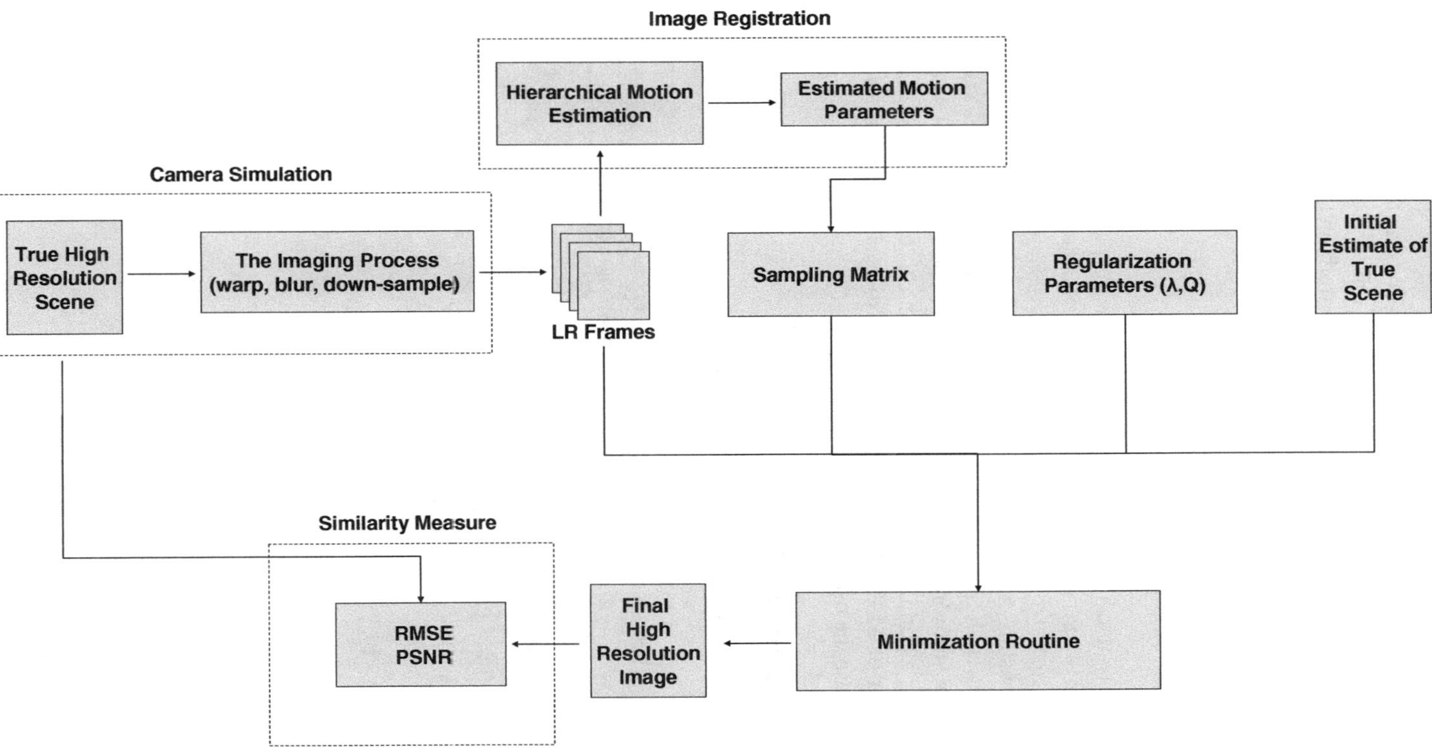

Fig. 6.4 Image Super-Resolution Optimization (ISRO) Block Diagram

- *Similarity Measure*: To evaluate the reconstruction quality of the final ISRO-generated high-resolution image, one could measure the similarity between the actual true scene or ground truth and the reconstructed scene. This would include calculating RMSE and PSNR. This module is only applicable when the original true high-resolution scene is known and available for comparison.

6.3.3 Software Tools

The ISRO system is a MATLAB-based software package, built on MATLAB 7.1.0.246 (R14) Service Pack 3 using Image Processing Toolbox v5.1. Due to the complexity and size of the problem, for the purpose of large-scale optimization, instead of using MATLAB's Optimization Toolbox, we employ TOMLAB development environment in MATLAB. TOMLAB v4.8 provides state-of-the-art optimization software packages which are faster and more robust as compared to MATLAB's optimization toolbox. The graphical user interface development environment of MATLAB, GUIDE, was used to create graphical user interfaces (GUIs) to ease the use of the ISRO system.

6.3.4 GUI Screenshots

Following are the screenshots of the ISRO-GUI system.

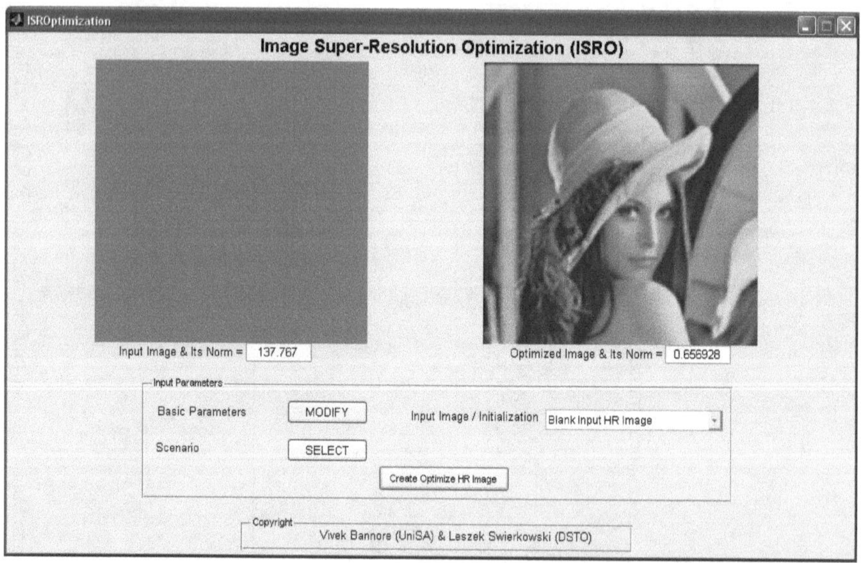

Fig. 6.5 ISRO Graphical User Interface

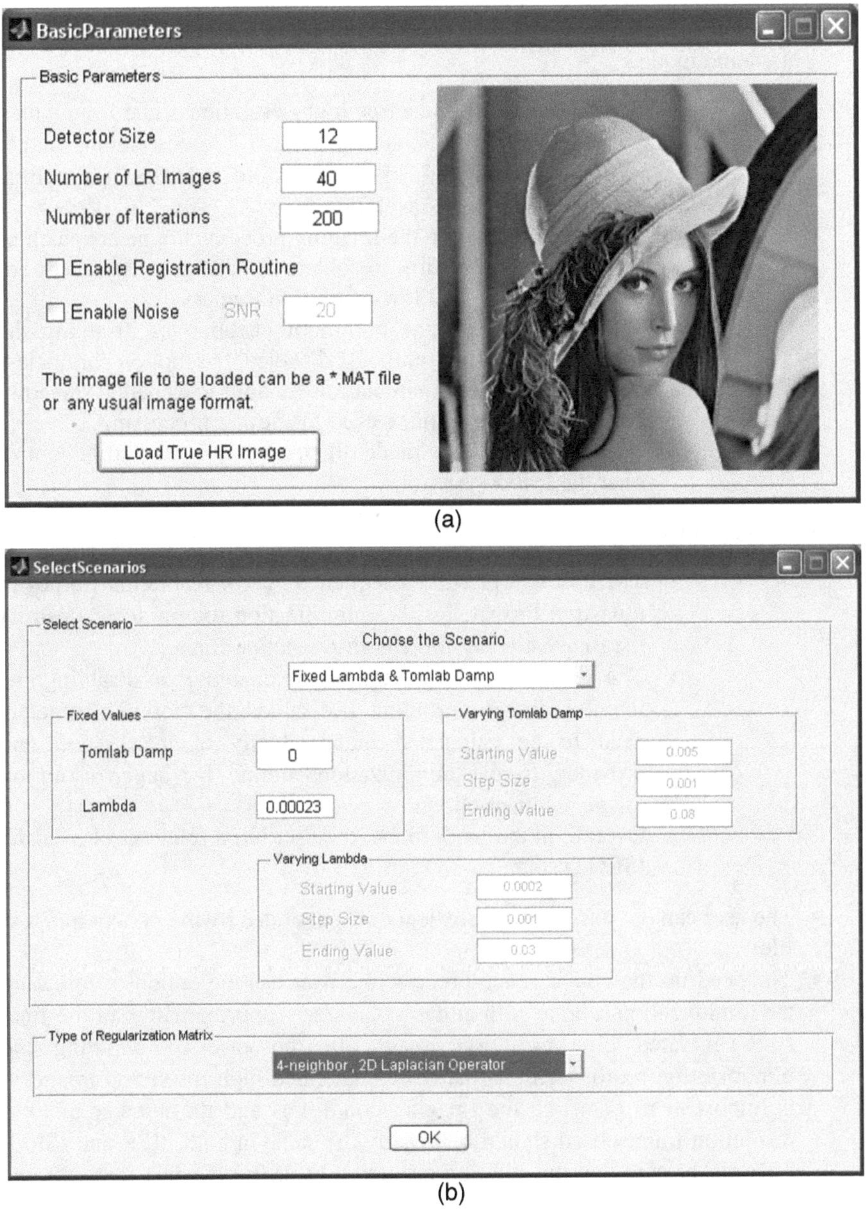

(a)

(b)

Fig. 6.6 GUIs called from within the main ISRO GUI. (a) Basic Parameters GUI, and (b) Select Scenario GUI.

6.3.5 Salient Features

Following are some of the most important features of the ISRO software:

- The software can be executed in either a demo mode or work mode.
- In demo mode,
    - The user may choose a true high-resolution scene from which a sequence of artificially generated low-resolution images will be created and will be used to reconstruct the final estimate of the high-resolution frame.
    - The user can change the imaging process parameters, such as the decimation ratio, to obtain a different sequence of artificially generated low-resolution frames.
    - The user also has the option of enabling or disabling the inbuilt registration routine. If disabled, the motion parameters used by the camera simulation module to generate the low-resolution frames will be used for further processing.
    - Overall, in the demo mode, the user only has a partial control over the ISRO system.
- In work mode,
    - The user can provide a sequence of low-resolution frames in a desired format, for the minimization routine to reconstruct the final estimate of the high-resolution frame.
    - The user also has the option of enabling or disabling the inbuilt registration routine. If disabled, the motion parameters need to be calculated independently of the system and provided to the minimization routine for super-resolution image reconstruction to proceed.
    - Overall, in the work mode, the user has a full control over the ISRO system.
- The user can opt to save the statistical data generated by the system in a text file.
- To speed up the convergence process, the user has the option of initiating the minimization routine with either a blank/zero approximation or the final IISR-generated high-resolution image. In the case of initiating the minimization routine with a final IISR-generated high-resolution image, it is important to note that the imaging conditions and the number of low-resolution frames used should be exactly the same in both, IISR and ISRO. This makes sure that the solution generated by ISRO is consistent with the type of initialization.
- For the purpose of understanding the key role of the regularization term in the reconstruction of a high-resolution image, the user has the option of experimenting with 4 different forms of the stabilization matrix, $\boldsymbol{Q}$.
- For the same purpose, the user also has the flexibility in selecting whether to execute the minimization routine for a specific value or over a range of values of the regularization parameter or damp factor, $\boldsymbol{\lambda}$.

6.4 Summary

During our research, MATLAB-based softwares were produced for solving the problem of super-resolution. The two major softwares that were produced are Iterative-Interpolation Super-Resolution (IISR) and Image Super-Resolution Optimization (ISRO). Both softwares can act as standalone super-resolution image reconstruction packages. They also serve the purpose of being experimental tools for understanding the effect on the fidelity of reconstruction by implementing different,

- interpolation kernels in the IISR technique, or
- regularization terms in ISRO package.

For the ease of understanding the working of these softwares, skeleton diagrams have been provided along with brief descriptions of each and every process involved. To achieve maximum flexibility, the softwares have been developed to be as modular as possible. GUIs have also been created for both the softwares to provided maximum usability.

Chapter 7
Conclusion and Future Directions

In this chapter we conclude by emphasizing the contributions of this research in the field of image super-resolution reconstruction. Possible future research directions for image super-resolution are also provided.

7.1 Contribution of This Research

This research addresses the problem of super-resolution image enhancement in terms of maintaining highest fidelity of reconstruction and a low computational cost to achieve maximum applicability of super-resolution to the real-world applications [80].

A novel and hybrid reconstruction scheme has been proposed for solving the problem of super-resolution restoration of high-resolution images from sequences of geometrically warped, aliased and under-sampled low-resolution images. This technique is known as the ***Iterative-Interpolation Super-Resolution*** (**IISR**). The proposed reconstruction scheme uses interpolation techniques to produce the *first approximation* of high-resolution image and then employs an iterative approach to generate the final solution. The extensive analytical data presented for the IISR system, illustrates the effectiveness and robustness of the proposed reconstruction scheme. As expected, the higher degree interpolation kernels were more accurate leading to better reconstruction but at a higher cost of computational time. The IISR scheme also contains implicit regularization features because both, the size of the interpolation kernel as well as the number of iterations, strongly affect the smoothness of the reconstructed HR image and controls the stability of the process.

For reinvestigating the influence of regularization term on the accuracy of super-resolution image reconstruction, several different forms of regularization terms were studied in this research. Detailed analysis based on the experimental data generated by using the different forms of regularization terms is also presented. In order to see the effect of the regularization parameter, λ, over the final solution, L-curve and λ-curve were considered. From our experiments, it was seen that super-resolution reconstruction was not very sensitive to the precise value of the regularization parameter when the parameter was larger than the true optimal value. This has practical implications and also explains why in many simulations good reconstructions were achieved with regularization parameter chosen in an ad hoc manner. However, because our simulations were limited only to several examples of imagery the observed insensitivity of the super-resolution reconstruction to the exact value of λ may not be a universal property.

V. Bannore: Iterative-Interpolation Super-Resolution Img. Recons., SCI 195, pp. 105–107.
springerlink.com

Taking the advantage of the detailed analysis of the regularization term, the accuracy and efficiency of the IISR technique is evaluated by comparing it against the optimization technique using the best possible regularization term. After conducting several computer experiments, it was observed that the IISR reconstruction was reasonably accurate and that further improvement by the optimization procedure was relatively small and computationally expensive. The analysis also illustrate that for a faster convergence of the optimization procedure, the final IISR solution should be used as an initialization or starting point instead of a blank input HR image.

To address the applicability of super-resolution techniques in a real world computer vision application, accurate alignment of all geometrically warped, aliased and under-sampled low-resolution frames is mandatory. A hierarchical architecture motion estimation technique based on [5] is adopted and implemented for accurate estimation of relative motion between the rotated and translationally shifted, aliased, under-sampled, low-resolution frames. Even in the case of noise contaminated low-resolution frames, hierarchical motion estimation scheme ensures accurate estimation of angle of rotation and translational shifts for precise alignment. Using a coarse-to-fine refinement strategy along with the pyramid architecture, the registration technique comparatively has a low computational cost. The technique therefore proves to be well-suited, to be incorporated with the proposed fast and efficient IISR scheme, to provide a complete super-resolution image reconstruction package.

It cannot be emphasized more, that for the implementation of super-resolution reconstruction technique in a real environment, how important it is to maintain a proper balance between improving spatial resolution and keeping the computational time low. The proposed IISR system requires a relatively small number of low-resolution images for efficient reconstruction. Good results were obtained with only 10 LR frames for magnification factors as large as 20. This is important for practical applications, because if a large number of low-resolution images were required the accumulation of errors would impede the reconstruction accuracy. In order to evaluate the robustness of the proposed IISR system, all computer simulations and analytical data presented in this research were produced using noiseless and noise contaminated, aliased and under-sampled, low-resolution frames. Also, since each iteration in the IISR scheme only requires simple operations the technique is fast and computationally efficient. With further development, the technique certainly is promisingly suitable for hardware implementation for real-time processing.

Lastly, two major professional software frameworks, IISR and ISRO, have been presented in this work based on our research. To achieve maximum flexibility and usability, the softwares have been made modular and GUIs have been created using MATLAB's GUIDE. Both softwares serve the purpose of being standalone super-resolution image reconstruction packages. They can also be utilized as experimental tools for comparing the accuracy of reconstruction by generating high-resolution images, using different interpolation kernels in the IISR package or different combinations of the regularization term, $\boldsymbol{\lambda}$ and $\boldsymbol{Q}$, in the ISRO package. With further development, both the software packages, IISR and ISRO, are commercially promising and valuable for further research and development of computationally efficient super-resolution image reconstruction techniques.

7.2 Future Directions

There are several areas of super-resolution in which future research may be performed. One such important area is, developing of a more sophisticated and elaborate motion model. Such a model could incorporate complex motions such as occlusion or non-linear motion. Estimating the motion parameters accurately in such a motion model is a challenging problem by itself. As the complexity of the problem along with the image data size increases, the computational speed of accurate motion estimation also increases. It is therefore important to develop registration techniques for complex motion models with comparatively low computation speed for applicability in real world computer vision applications.

Another area of research would be to develop a fast object tracking algorithm. This would enable a user to select a ROI in the low-resolution frames and the tracking algorithm would have to intelligently track the object through all the frames. It would be advantageous to specify a threshold or minimum area of ROI to be tracked so that any unwanted erroneous pixels do not get selected into the ROI after tracking. An extension of this could be automating the selection of ROI and tracking it, that is selecting, identifying and tracking possible targets within a sequence of low-resolution frames. For example, selecting a car's rear license plate or intelligently marking the boundary of a human fig. or face in the center of the frame as the ROI.

Acquired images are generally contaminated by noise due to the imperfect imaging systems. Noise can also be introduced due to compression or transmission errors. In the last two decades a number of denoising techniques have been proposed in the literature. Among these, techniques based on wavelets have gained more popularity as compared to the ones based on spatial and Fourier domains. Developing a fast and robust denoising technique based on wavelet transforms for large-scale problems is a promising future direction for research as well.

References

[1] Photography - Electronic still-picture cameras - Resolution measurements, ISO 12233:2000, International Standard Organization (2000)

[2] Nyquist, H.: Certain topics in Telegraph Transmission Theory. Trans. Amer. Inst. Elect. Eng. 47, 617–644 (1928)

[3] Shannon, C.E.: A mathematical theory of communication. The Bell System Technical Journal 27, 379–423 (1948)

[4] Shannon, C.E.: Communication in the Presence of Noise. Proc. IEEE 86(2), 447–457 (1998)

[5] Bergen, J.R., Anandan, P., Hanna, K.J., Hingorani, R.: Hierarchical Model-Based Motion Estimation. In: Proceedings of the Second European Conference on Computer Vision, pp. 237–252. Springer, Heidelberg (1992)

[6] Alam, M.S., Bognar, J.G., Hardie, R.C., Yasuda, B.J.: High-Resolution Infrared Image Reconstruction Using Multiple Randomly Shifted Low-Resolution Aliased Frames. In: Infrared Imaging Systems: Design, Analysis, Modelling, and Testing VIII, SPIE Proceedings, vol. 3063 (April 1997)

[7] Elad, M., Feuer, A.: Restoration of a Single SR Image from Several Blurred, Noisy & Under-sampled Measured Images. IEEE Trans. on Image Processing 6, 1646–1658 (1997)

[8] Chaudhuri, S. (ed.): Super-Resolution Imaging, 1st edn., p. 279. Kluwer Academic Publishers, Dordrecht (2001)

[9] Kang, M.G., Chaudhuri, S.: Super-Resolution Image Reconstruction. IEEE Signal Processing Magazine 20, 19–20 (2003)

[10] Farsiu, S., Robinson, D., Elad, M., Milanfar, P.: Advances and Challenges in Super-Resolution. International Journal of Imaging Systems and Technology 14, 47–57 (2004)

[11] Tsai, R.Y., Huang, T.S.: Multiframe Image Restoration and Registration, vol. 1, pp. 317–339. JAI Press Inc., Greenwich (1984)

[12] Kim, S.P., Bose, N.K., Valenzuela, H.M.: Recursive Reconstruction of High-Resolution Image from Noisy Undersampled Multiframes. IEEE Trans. on Acoustics, Speech, Signal Processing 38, 1013–1027 (1990)

[13] Kim, S.P., Su, W.Y.: Recursive High-Resolution Reconstruction of Blurred Multiframe Images. IEEE Trans. on Image Processing 2, 534–539 (1993)

[14] Irani, M., Peleg, S.: Improving Resolution by Image Registration. CVGIP: Graphic Models and Image Processing 53, 231–239 (1991)

[15] Irani, M., Peleg, S.: Motion Analysis for Image Enhancement: Resolution, Occlusion & Transparency. Journal of Visual Communication and Image Representation 4, 324–335 (1993)
[16] Hardie, R.C., Bognar, J.G., Barnard, K.J., Watson, E.A.: High-Resolution Image Reconstruction from a Sequence of Rotated and Translated Frames and It's Application to an Infrared Imaging System. Optical Engineering 37, 247–260 (1998)
[17] Alam, M.S., Bognar, J.G., Hardie, R.C., Yasuda, B.J.: Infrared image registration and high-resolution reconstructionusing multiple translationally shifted aliased video frames. Instrumentation and Measurement 49(5), 915–923 (2000)
[18] Tuinstra, T.R., Hardie, R.C.: High-resolution image reconstruction from digital video by exploitation of non-global motion. Optical Engineering 38, 806–814 (1999)
[19] Shen, H., Li, P., Zhang, L., Zhao, Y.: A MAP algorithm to super-resolution image reconstruction. In: Proceedings of Third International Conference on Image and Graphics, pp. 544–547 (December 2004)
[20] Tom, B.C., Katsaggelos, A.K., Galatsanos, N.P.: Reconstruction Of A High Resolution Image From Registration And Restoration Of Low Resolution Images. In: Proc. IEEE Int. Conf. Image Processing, Austin, TX, vol. 3, pp. 553–557 (1994)
[21] Tom, B.C., Katsaggelos, A.K., Galatsanos, N.P.: Reconstruction of a High Resolution Image by Simultaneous Registration, Restoration, and Interpolation of Low-Resolution Images. In: IEEE International Conference on Image Processing, vol. 2, pp. 539–542 (October 1995)
[22] Youla, D.C., Webb, H.: Image Restoration by the method of Convex Projections: Part 1-Theory. IEEE Trans. on Medical Imaging MI-1(2), 81–94 (1982)
[23] Sezan, M.I., Stark, H.: Image Restoration by the method of Convex Projections: Part 2-Applications and Numerical Results. IEEE Trans. on Medical Imaging MI-1(2), 95–101 (1982)
[24] Stark, H.: Theory of Convex Projection and Its Application to Image Restoration. In: IEEE International Symposium on Circuits & Systems, vol. 1, pp. 963–964 (June 1988)
[25] Stark, H.: Convex Projections in Image Processing. In: IEEE International Symposium on Circuits & Systems, vol. 3, pp. 2034–2036 (May 1990)
[26] Stark, H., Oskoui, P.: High-resolution image recovery from image-plane arrays, using convex projections. Optical Society of America, Journal A 6, 1715–1726 (1989)
[27] Caner, G., Tekalp, A.M., Heinzelman, W.: Super Resolution Recovery for Multi-Camera Surveillance Imaging. In: Proceedings of International Conference on Multimedia and Expo, ICME 2003, vol. 1, pp. I - 109–112 (July 2003)
[28] Nguyen, N., Golub, G., Milanfar, P.: Preconditioners for Regularized Image Superresolution. In: Proceedings of IEEE International Conference on Acoustics, Speech, and Signal Processing, 1999. ICASSP 1999, vol. 6, pp. 3249–3252 (March 1999)
[29] Nguyen, N., Milanfar, P., Golub, G.: A Computationally Efficient Super-Resolution Image Reconstruction Algorithm. IEEE Transactions on Image Processing 10(4), 573–583 (2001)
[30] Elad, M., Feuer, A.: Super-Resolution Restoration Of An Image Sequence: Adaptive Filtering Approach. IEEE Transactions on Image Processing 8, 387–395 (1999)

[31] Elad, M., Feuer, A.: Super-Resolution Reconstruction of Continuous Image Sequences. In: Proceedings of the 1999 International Conference on Image Processing (ICIP 1999), vol. 3, pp. 459–463 (1999)

[32] Elad, M., Feuer, A.: Super-resolution reconstruction of image sequences. IEEE Transactions on Pattern Analysis and Machine Intelligence 21(9), 817–834 (1999)

[33] Freeman, W.T., Pasztor, E.C., Carmichael, O.T.: Learning Low-Level Vision. International Journal of Computer Vision 40(1), 25–47 (2000)

[34] Baker, S., Kanade, T.: Limits on Super-Resolution and How to Break Them. IEEE Trans. Pattern Anal. and Machine Intell. 24(9), 1167–1183 (2002)

[35] Freeman, W.T., Jones, T.R., Pasztor, E.C.: Example-Based Super-Resolution. IEEE Computer Graphics and Applications 22(2), 56–65 (2002)

[36] Chang, H., Yeung, D.Y., Xiong, Y.: Super-Resolution Through Neighbor Embedding. In: IEEE Computer Society Conference on Computer Vision and Pattern Recognition, vol. 1, pp. 275–282 (July 2004)

[37] Qiao, J., Liu, J., Chen, Y.-W.: Joint Blind Super-Resolution and Shadow Removing. IEICE Transactions on Information and Systems E90-D(12), 2060–2069 (2007)

[38] Whittaker, E.T.: On the functions which are represented by the expansion of interpolating theory. Proc. R. Soc. Edinburgh 35, 181–194 (1915)

[39] Jerri, A.J.: The Shannon Sampling Theorem—Its Various Extensions and Applications: A Tutorial Review. Proc. IEEE 65, 1565–1595 (1977)

[40] Keys, R.: Cubic Convolution Interpolation for Digital Image Processing. IEEE Transactions on Acoustics, Speech, and Signal Processing [see also IEEE Transactions on Signal Processing] 29(6), 1153–1160 (1981)

[41] Maeland, E.: On the Comparison of Interpolation Methods. IEEE Transactions on Medical Imaging 7(3), 213–217 (1988)

[42] Appledorn, C.R.: A New Approach to the Interpolation of Sampled Data. IEEE Trans. on Medical Imaging 15(3), 369–376 (1996)

[43] Press, W.H., Flannery, B.P., Teukolsky, S.A., Vetterling, W.T.: Iterative Improvement of a Solution to Linear Equations. In: Numerical Recipes in C, 2nd edn., October 1992, p. 1020. Cambridge University Press, Cambridge (1992)

[44] Bannore, V., Swierkowski, L.: An Iterative Approach to Image Super-Resolution. In: Shi, Z., K., S., D., F. (eds.) Intelligent Information Processing III, vol. 228, pp. 473–482. Springer, Boston (2006)

[45] Bannore, V., Swierkowski, L.: Fast Iterative Super-Resolution for Image Sequences. In: 9th Biennial Conference of the Australian Pattern Recognition Society on Digital Image Computing Techniques and Applications (DICTA 2007), Adelaide, Australia, pp. 286–293. IEEE - Computer Society, Los Alamitos (2007)

[46] Bannore, V., Swierkowski, L.: Search for a computationally efficient image super-resolution algorithm. In: Apolloni, B., Howlett, R.J., Jain, L. (eds.) KES 2007, Part I. LNCS, vol. 4692, pp. 510–517. Springer, Heidelberg (2007)

[47] Hadamard, J.: Sur les problèmes aux dérivées partielles et leur signification physique (On the problems with the derivative partial and their physical significance), pp. 49–52. Princeton University Bulletin (1902)

[48] Groetsch, C.W.: The Theory of Tikhonov Regulaization for Fredholm Equations of the First Kind. Research Notes in Mathematics, vol. 105. Pitman, Boston (1984)

[49] Engl, H.W.: Regularization methods for the stable solution of Inverse Problems. Surveys on Mathematics for Industries 3, 77–143 (1993)

[50] Hansen, P.C.: Analysis of Discrete Ill-Posed Problems by means of the L-Curve. Siam Review 34(4), 561–580 (1992)

[51] Galatsanos, N.P., Katsaggelos, A.K.: Methods for choosing the Regularization Parameter and Estimating the Noise Variance in Image Restoration and Their Relation. IEEE Transactions on Image Processing 1(3), 322–336 (1992)

[52] Hanke, M., Hansen, P.C.: Regularization Methods For Large-Scale Problems. Surveys on Mathematics for Industries 3, 253–315 (1993)

[53] Bannore, V.: Regularization for Super-Resolution Image Reconstruction. In: Gabrys, B., Howlett, R.J., Jain, L.C. (eds.) KES 2006. LNCS (LNAI), vol. 4252, pp. 36–46. Springer, Heidelberg (2006)

[54] Morozov, V.A.: On the Solution of Functional Equations by the method of Regularization. Soviet Math. Dokl. 7, 414–417 (1966)

[55] Golub, G.H., Heath, M.T., Wahba, G.: Generalized Cross-Validation as a method for choosing a good Ridge Parameter. Technometrics 21, 215–223 (1979)

[56] Wahba, G.: Spline Model for Observational Data, vol. 59. Society for Industrial and Applied Mathematics, Philadelphia (1990)

[57] Varah, J.M.: Pitfalls in the Numerical Solution of Linear Ill-Posed Problems. SIAM J. Sci. Stat. Comput. 4(2), 164–176 (1983)

[58] Hansen, P.C., O'Leary, D.P.: The Use of the L-Curve in the Regularization of Discrete Ill-Posed Problems. Siam J. Sci. Comput. 14(6), 1487–1503 (1993)

[59] Lawson, C.L., Hanson, R.J.: Solving least squares problems. Prentice Hall, Englewood Cliffs (1974)

[60] Tikhonov, A.N.: Regularization of Incorrectly Posed Problems. Soviet Math. Dokl. 4, 1624–1627 (1963)

[61] Tikhonov, A.N.: Solution of Incorrectly Formulated Problems and the Regularization Method. Soviet Math. Dokl. 4, 1035–1038 (1963)

[62] Tikhonov, A.N.: Ill-Posed Problems in Natural Sciences. TVP, Sci. Publishers, Moscow (1991)

[63] Zhuang, X., Ostevold, E., Haralick, M.: A Differential Equation Approach To Maximum Entropy Image Reconstruction. IEEE Trans. Acoust., Speech, Signal Processing ASSP-35(2), 208–218 (1987)

[64] Hansen, P.C.: Rank-Deficient and Discrete Ill-Posed Problems: Numerical Aspects of Linear Inversion. SIAM, Philadelphia (1998)

[65] Brown, L.G.: A Survey of Image Registration Techniques. ACM Computing Surveys 24(4), 325–376 (1992)

[66] Zitova, B., Flusser, J.: Image Registration Methods: A Survey. IVC 21(11), 977–1000 (2003)

[67] Zhao, W., Sawhney, H.S.: Is super-resolution with optical flow feasible? In: Heyden, A., Sparr, G., Nielsen, M., Johansen, P. (eds.) ECCV 2002. LNCS, vol. 2350, pp. 599–613. Springer, Heidelberg (2002)

[68] Aggarwal, J.K., Nandhakumar, N.: On the Computation of Motion from Sequences of Images-A Review. Proceedings of the IEEE 76(8), 917–935 (1988)

[69] Horn, B.K.P., Schunck, B.G.: Determining Optical Flow. Artificial Intelligence 17, 185–203 (1981)

[70] Lin, T., Barron, J.: Image Reconstruction Error for Optical Flow. In: Vision Interface, Banff National Park, Alberta, Canada, pp. 73–80 (May 1994)

[71] Baker, S., Kanade, T.: Super-Resolution Optical Flow. The Robotics Institute, Carnegie Mellon University (October 1999)

[72] Hardie, R.C., Barnard, K.J., Armstrong, E.E.: Joint MAP Registration and High-Resolution Image Estimation Using a Sequence of Undersampled Images. IEEE Transactions on Image Processing 6(12), 1621–1633 (1997)

[73] Unser, M., Aldroubi, A., Gerfen, C.R.: A Multiresolution Image Registration Procedure Using Spline Pyramids. In: Proceedings of the SPIE Conference on Mathematical Imaging: Wavelet Applications in Signal and Image Processing, vol. 2034, pp. 160–170 (1993)

[74] Sharma, R.K., Pavel, M.: Multisensor Image Registration. Society for Information Display XXVIII, 951–954 (1997)

[75] Thevenaz, P., Ruttimann, U.E., Unser, M.: A Pyramid Approach to Subpixel Registration Based on Intensity. IEEE Transactions on Image Processing 7(1), 27–41 (1998)

[76] Burt, P.J., Adelson, E.H.: The Laplacian Pyramid as a Compact Image Code. IEEE Transactions on Communications Com-31(4), 532–540 (1983)

[77] Adelson, E.H., Anderson, C.H., Bergen, J.R., Burt, P.J., Ogden, J.M.: Pyramid Methods in Image Processing. RCA Engineer 29(6), 33–41 (1984)

[78] Unser, M., Aldroubi, A., Eden, M.: B-Spline Signal Processing: Part I - Theory. IEEE Transactions on Signal Processing 41(2), 821–833 (1993)

[79] Unser, M., Aldroubi, A., Eden, M.: B-Spline Signal Processing: Part II - Efficient Design and Applications. IEEE Transactions on Signal Processing 41(2), 834–848 (1993)

[80] Bannore, V.: Computationally Efficient Iterative-Interpolation Super-Resolution Image Reconstruction Scheme. University of South Australia (2008)

Zeitfracht Medien GmbH
Ferdinand-Jühlke-Straße 7
99095 Erfurt, Deutschland
produktsicherheit@kolibri360.de